SUSTAINING EARTH: RESPONSE TO THE ENVIRONMENTAL THREAT

Sustaining Earth:

Response to the Environmental Threat

Edited by
D.J.R. Angell, J.D. Comer
and
M.L.N. Wilkinson

M
MACMILLAN

First published 1990

Published by
MACMILLAN ACADEMIC AND PROFESSIONAL LTD
Houndmills, Basingstoke, Hampshire RG21 2XS
and London
Companies and representatives
throughout the world

Typeset by Megaron, Cardiff Wales

Printed in Great Britain by
WBC Ltd., Bridgend

British Library Cataloguing in Publication Data
Sustaining Earth: response to the environmental threat.
1. Environment. Research
I. Angell, D.J.R. II. Comer, J.D. III. Wilkinson, M.L.N.
333.7072

ISBN 0-333-52492-6

To Kate, Pabs, and Leslie

Contents

List of Tables

Acknowledgements

The editors wish to acknowledge a number of debts. For their assistance, we are grateful to Dr Richard Grove of Churchill College, Cambridge; Richard Langhorne, of the Centre of International Studies, Cambridge University; Michael McCarthy, of *The Times*; Tessa Murray and the Government and Public Affairs Department of British Petroleum Company plc; Sir Arthur Norman and David Cope of the UK Centre for Economic and Environmental Development; Richard Sandbrook and Czech Conroy, of the International Institute for Environment and Development; Sir William Wilkinson of the Nature Conservancy Council; and Nigel Haigh of the Institute for European Environmental Policy. For their support, we would like to thank our families and friends.

List of Abbreviations

ASEAN	Association of South East Asian Nations
BAS	British Antarctic Survey
BCC	British Council of Churches
CAFOD	Catholic Fund for Overseas Development
CAP	Common Agricultural Policy (Europe)
CFC	chlorofluorocarbon
CIDA	Canadian International Development Agency
CSCE	Conference on Security and Cooperation in Europe
EEC/EC	European Economic Community/European Community
FAO	Food and Agricultural Organisation of the United Nations
GNP	gross national product
G77	Group of 77
HMIP	Her Majesty's Inspectorate of Pollution (UK)
IIED	International Institute for Environment and Development (UK)
IRPP	Institute for Research on Public Policy (Canada)
IUCN	International Union for Conservation of Nature and Natural Resources
NAS	National Academy of Sciences (USA)
NASA	National Aeronautics and Space Administration (USA)
NGO	non-governmental organisation
NRC	National Research Council (Canada)
ODA	Overseas Development Administration (UK)
OECD	Organisation for Economic Co-operation and Development
RCEP	Royal Commission on Environmental Pollution (UK)
RSA	Royal Society of Arts (UK)
UNEP	United Nations Environment Programme
UNESCO	United Nations Educational, Scientific and Cultural Organisation
UNFPA	United Nations Fund for Population Activities
USAID	United States Agency for International Development
WCC	World Council of Churches
WCED	World Commission on Environment and Development
WCS	World Conservation Strategy
WWF	World Wide Fund for Nature (formerly the World Wildlife Fund)

Notes on the Contributors

David Angell is a member of St John's College, Cambridge.

Dr Gro Harlem Brundtland is Chairman of the World Commission on Environment and Development and was Prime Minister of Norway in 1981 and from 1986 to 1989. She was Minister of the Environment between 1974 and 1979 and was a member of the Independent (Palme) Commission on Disarmament and Security Issues.

Hon Charles Caccia, PC, MP, was Minister for the Environment in Canada under Prime Minister Pierre Trudeau. He has been a Member of Parliament since 1968 and is currently head of the Parliamentary Centre for Sustainable Development.

Stanley Clinton Davis was European Commissioner for the Environment on the Commission of the European Communities between 1985 and 1988. Previously, he was a Member of Parliament in Britain and served as Parliamentary Under-Secretary of State for Trade and Opposition Spokesman on Trade and Foreign Affairs.

Justyn Comer is a member of Trinity College, Cambridge.

Joe Farman has been a member of the British Antarctic Survey (BAS), Cambridge, since 1956. He led the team of scientists which first produced evidence of ozone depletion in the upper atmosphere, in the journal *Nature* in 1985.

David Gosling was Director of Church and Society at the World Council of Churches. He is currently at Great St Mary's, the University Church, Cambridge.

Dr Richard Grove is a Fellow of Churchill College, Cambridge.

Christopher Hampson is Executive Director of Imperial Chemical Industries plc.

Dr Martin Holdgate, CB, is Director-General of the International Union for Conservation of Nature and Natural Resources, in Gland, Switzerland. Previously, he was Chief Scientist and Deputy Secretary

(Environmental Protection) at the Department of the Environment in Britain and, in 1983–4, President of the Governing Council of the United Nations Environment Programme (UNEP).

Jim MacNeill is President of MacNeill Associates and Director of the Environment and Sustainable Development Programme at The Institute for Research on Public Policy, in Ottawa. He was a member (*ex-officio*) and Secretary-General of the World Commission on Environment and Development, and served previously as Director of Environment at the Organisation for Economic Co-operation and Development (OECD) for six years; Secretary (Deputy Minister) of the Canadian Ministry of State for Urban Affairs; Special Advisor on the Constitution and the Environment in Prime Minister Pierre Trudeau's Office; and Canadian Commissioner-General for the 1975 United Nations Conference on Human Settlements, in Vancouver.

Sir Arthur Norman, KBE, DFC, is Chairman of the UK Centre for Economic and Environmental Development (UKCEED) and Chairman of Trustees of the World Wide Fund for Nature UK (WWF UK). He was Chairman of the International Institute for Environment and Development (IIED), and of the De La Rue Company plc; President of the Confederation of British Industry; and a member of the Nature Conservancy Council.

Rt Hon Christopher Patten, PC, MP, is Secretary of State for the Environment in the United Kingdom. He was Minister of State (Foreign and Commonwealth Office) and Minister of Overseas Development when he delivered his speech for the Cambridge Lectures on Environment and Development.

Javier Perez de Cuellar is Secretary-General of the United Nations. Previously, he was Peru's Ambassador to Switzerland, the Soviet Union, Venezuela and the United Nations, and Permanent Under-Secretary and Secretary General of the Peruvian Foreign Office.

Professor Ghillean Prance has been Director of the Royal Botanic Gardens, Kew, England, since 1988. Previously he was a member of the New York Botanical Gardens from 1963, and was Senior Vice President for Science from 1981. He has travelled extensively throughout the Amazon region carrying out a botanical survey of Brazilian Amazonia.

Jules Pretty is Associate Director of the Sustainable Agriculture Programme at the International Institute for Environment and Development, London.

Sir Shridath S. Ramphal, OE, AC, CMG, QC, SC, was Commonwealth Secretary-General between 1975 and 1990. He is a member of the World Commission on Environment and Development and of the South Commission which is exploring North/South issues and looking at enhanced South/South co-operation. Previously, he was Attorney-General, Minister of Justice and Minister of Foreign Affairs in Guyana. Sir Shridath was a member of the Independent (Brandt) Commission on International Development Issues, the Independent (Palme) Commission on Disarmament and Security Issues, and the Independent (Aga Khan) Commission on International Humanitarian Issues. His chapter is based on a speech delivered as part of the Cambridge Lectures on Environment and Development.

Dr Stephen H. Schneider is head of the interdisciplinary climate-systems section at the National Science Foundation's National Center for Atmospheric Research at Boulder, Colorado.

Sir Crispin Tickell, GCMG, KCVO, is Ambassador and Permanent Representative of the United Kingdom to the United Nations. He was Permanent Secretary of the Overseas Development Administration (ODA) from 1984 to 1987, Ambassador to Mexico from 1981 to 1983, Visiting Fellow of All Souls, Oxford, in 1981, and Chef de Cabinet to the President of the Commission of the European Communities from 1977 to 1981. Sir Crispin is the author of *Climatic Change and World Affairs* (1977, 1986).

Matthew Wilkinson is a member of Trinity College, Cambridge.

Dr Sarah Woodin is Chief Scientist on Air Pollution at the Nature Conservancy Council, Peterborough, England.

Foreword

Gro Harlem Brundtland

The unknown quality of the future has always stimulated man's creativity. But at the same time it has inspired fear and uncertainty. Can we cope with the challenges, do we know where we are going, can we choose the right direction?

Every generation throughout history has reflected on these uncertainties. Yet the challenges we are facing towards the end of this century are without precedent. For the first time in history, millions of people all over the world are not only worried about their own and their children's future; they are deeply concerned about the future of the entire planet.

They are right to be anxious. Scientists are drawing our attention to urgent problems bearing on humanity's very survival: global warming, threats to the Earth's ozone layer, deserts consuming agricultural land.

These are challenges of a new kind. Most of them are outside our day-to-day concerns, but the growing evidence leave no doubt that they are real, that they affect each and every one of us. There is growing awareness that they must be met if we are to avoid a steady deterioration in our living conditions.

This awareness has, however, been matched by a growing frustration about our own ability to address these vital issues and deal with them effectively. We see that there is a need for change, but we do not have the answers at hand. We know that action is called for, but national and international institutions seem unable to arrive at the right form of action.

This was the background against which the UN General Assembly set up the World Commission on Environment and Development in 1983. It was a logical successor to two other commissions which called for international political action: after the Brandt Commission's Programme for Survival and Common Crisis and the Palme Commission on Common Security, the time had come to address the need to secure 'Our Common Future'.

Introduction

D.J.R. Angell, J.D. Comer, M.L.N. Wilkinson

The World Commission on Environment and Development was established in 1983 at the instigation of the United Nations and under the chairmanship of Dr Gro Harlem Brundtland of Norway. Its mandate was to explore the nature and possible consequences of the environmental threats that face mankind and to recommend measures to safeguard and improve the quality of life on earth in the future, taking into account the interrelationships between people, resources, environment and development. The Commission concluded that provision for an acceptable future could only be made by the adoption of sustainable development as the working principle behind all future planning. Sustainable development was defined as: 'development that meets the needs of the present without compromising the ability of future generations to meet their own needs'.

Thorough and visionary though the Report is, the holistic perceptions of one earth that were its essence are far from realised. It remains a gauntlet thrown down for mankind to pick up and act upon, or to leave alone, hoping with our heads in the sand that the problems will go away.

In 1988, a year after the Commission's Report was published, we organised a series of seven public lectures at Cambridge University to examine how politicians, industrialists, scientists and the public were responding to the Report's recommendations and to the degradation that it highlighted. This book is a continuation of the lectures and their theme. It is a compilation of responses that represent a wide variety of attitudes towards environmental matters and the idea of sustainable development.

We do not intend or hope to offer a thorough environmental education. Rather, we hope to inform the reader to the extent that he or she may enter into the debate on environmentalism and conservation with a surer feel for the arguments being championed. There is therefore no consistent thesis to the book, only the constant theme of the response to the challenge of sustainable development. We also hope that you will encounter sections or sentences that encourage you to probe beyond the ideas presented here, for there is no doubt that many of the environmental dictums that now stand largely unchallenged will soon be opposed. A view expressed by a number of contributors, for example, is that birth control, hand-in-hand with the alleviation of poverty, will be an essential measure for stemming environmental degradation in the developing world. However, as the Columbian writer Gabriel García Márquez has observed, many Latin Americans regard their swelling numbers as the means by which they will in future challenge the power centres of North America and Europe. Those who are sufficiently angered by the unjust distribution of wealth and power may tolerate very harsh living conditions in the hope that they may one day put things to rights.

A broader question that must be beaten out in academic circles (but that we've also heard being discussed in the pub), is, 'Why should we save the planet?' Why should we attempt to achieve a sustainable future? Why should we conserve the environment for future generations? This question becomes more telling still when the overwhelming scale and complexity of the task of becoming sustainable is considered. Whether you appeal to the responsibilities of parenthood, the prevention of unforeseen suffering, disease and premature death or the much-touted 'enlightened self-interest', the word 'should' in the question and the morality that it implies are unavoidable. It may be that the environment will only receive the support it needs, in the face of conflicting interests and conveniences, if it is regarded as of sacred worth and in itself good and beautiful. It may also be that the sense of the sacred value of nature is still so innate in humans that it need very rarely be stated openly. It would be sad if the current surge of environmental concern subsided because the protection of nature could not withstand cynical or sceptical criticism. It would also be disastrous if continued reference to the purely instrumental value of nature for mankind proved able only to provoke an inadequate response from those whose livelihoods are not immediately threatened by environmental deterioration.

Predictions of the consequences of environmental change and of the timescales involved have in the past been far from reliable. The next decade will see much environmental questioning and an increase in research, possibly without the reward of many unambiguous answers. But the potential of the environmental threats of which we are aware is, without doubt, terrible. The urgency of the situation is such that in some instances preventative action will have to be taken, even though absolutely conclusive evidence is not yet available: such action may have to be carried out even if it poses a challenge to ideas and values which have become cultural commonplaces.

D.A., J.C., M.W.
Cambridge 1990

Part One: The Background

1 Endangered Earth

Sir Shridath S. Ramphal

Back in 1983, when Mrs Brundtland invited me to join the World Commission on Environment and Development, the world's environment was well down the list of priority political issues almost everywhere; academic interest in ecology was often treated by practitioners of more traditional disciplines with suspicion and disdain; as for 'environment and development', that was a specialism within a dubious specialism or, as Churchill might have said, a mystery wrapped in an enigma. Even environmentalists themselves sometimes seemed less concerned about people than pandas. I hesitated before accepting the Prime Minister's invitation. If the world wasn't ready for 'Brandt' on development, would it be ready for 'Brundtland' on environment and development?

Now – barely five years later – such hesitation would seem strange. The interrelated issues of environment and development vie with nuclear disarmament as the dominant issue of our time. Politicians, from Mr Gorbachev to Mrs Thatcher, and financiers, from the President of the World Bank to environmentally 'clean' unit trust managers, advertise their 'green' credentials. Britain – one of the few countries not to contribute to the costs of the Commission's work – decided to host the presentation of its Report; and hosted an important international conference on the ozone layer in March 1989. The *National Geographic* carried on its December 1988 cover a holograph of the planet which fractures as it tilts. *Time* magazine, at the start of 1989, dropped for only the second time in sixty years its 'Man of the Year' theme to emphasise that 1989 was to be the year of 'endangered earth'.

How has this transformation in perception come about? I like to think that the World Commission, and its Report *Our Common Future*, had something to do with it. But, perhaps more important was the way in which a succession of disasters all over the world triggered intellectual awareness about the possibility of some underlying pattern of causality, and aroused those emotions of fear and anger that are often the mainspring of political action. While the Commission met, some of those happenings graphically illustrated the dangers faced by humanity: Bhopal, Chernobyl, the Rhine chemical spillage, the mud slide in Colombia, the Mexican liquid gas explosion, drought and famine in Africa. When Commonwealth leaders debated the work of the World Commission at their meeting in Vancouver in 1987, the whole discussion was given immediacy by the recent experience of disastrous flooding in Bangladesh (worse was to come) and the inundation of the low-lying islands of the Maldives by unprecedented waves. Reinforced by empirical evidence that disasters have steadily increased in frequency in recent decades, political leaders were beginning to accept that all these – and more – are not purely random events.

But, while this spate of disasters has raised public and political consciousness about environmental stress, it is the quieter, less immediately dramatic trends which are, in many respects, more disturbing. The most recent estimates suggest that 11 million hectares of tropical forest – an area the size of East Germany – are being lost every year, mainly to land-clearing for crops and cattle ranching.[1] In India, where there tends to be a more honest and open discussion of environmental issues than elsewhere, it is now publicly acknowledged that forest loss is far more serious than previously recognised and, on present trends, little if any of the remaining 30 million hectares of forest will be left by the end of the century.[2] In Brazil, one of the few remaining tropical forest areas of any size, destruction proceeds apace. Chico Mendes, the rubber tappers' leader who fought, using non-violent Ghandian methods, to preserve Amazonia (and whose Association, incidentally, gave eloquent testimony to the Brundtland Commission when it visited Brazil) was murdered in December 1988 because he stood in the way of powerful interests wanting to destroy the forest. The costs of this forest destruction are only just beginning to be understood: long-term soil erosion; substantial contribution to the accumulation of greenhouse gases in the atmosphere; and irrevocable loss of plant and animal species which are becoming extinct at a rate of, perhaps, hundreds or even thousands of species a year.[3]

Deforestation is only one symptom of the declining health of the earth:[4] 6 million hectares of new desert – an area almost the size of the Irish Republic or Holland and Belgium combined – form annually;

thousands of lakes and rivers are biologically dead or dying; there is growing toxic contamination of water and soils. In India, 130 million hectares – almost 40 per cent of that country's land area – have been classified as degraded beyond the point of productive use. In Poland, one quarter of the soil is regarded as contaminated beyond the point of safe use and only 1 per cent of fresh water is now considered safe for drinking. In Mexico City, residents have been advised to jog indoors because of the dangers of breathing the air. Even in England's green and pleasant land, the claims of bottled water and organic farming no longer contend in vain.

For much of humanity, however, an even more pressing issue than where to jog or how to farm is how not to starve – how to rise above absolute poverty. Perhaps the most distinctive contribution of the Brundtland Commission was to provide a clear explanatory link between Third World poverty and global environmental deterioration – between economy and ecology. Traditionally, pollution has been seen largely as a by-product of wasteful life-styles and harmful production processes in the rich world. And in many respects that remains the case. 80 per cent of all commercial energy is generated in the industrialised world, including Eastern Europe; almost all the world's chlorofluorocarbons, the CFCs – those gases that are helping to destroy the ozone layer – originate in rich countries. But, in other respects, poverty is as degrading to the environment as it is to humans. The imperatives of daily survival force poor families to think (and live) short-term: to overgraze grasslands, to overexploit soils to maximise immediate yields, to cut down dwindling forest stocks for farmland or firewood. What is, individually, rational behaviour becomes collective folly.

Nowhere is there a wider disparity between understandable human choice and inevitable human disaster than in relation to population. For an individual family on the brink of survival it makes eminent sense to have large numbers of children, in the hope that some survive and help around the family farm or find work to support parents, brothers and sisters. But when many families do the same the combined result is to produce far more people than the stock of available fertile land and the infrastructure of schools, health and other services can sustain. In Kenya, a country that suffers acutely from land hunger, urban unemployment and environmental stress, a woman now produces eight children on average and the population is expected to rise from 25 to over 80 million in the next thirty-five years, even if the birth rate halves over that period. In Bangladesh, where almost every last acre of cultivatable land is already used – and millions live precariously on mudbanks facing imminent disaster – the population is expected to double from 110 million to 220 million over the same period of thirty-five years and, again, assuming a halving of the birth rate.[5] These are the pressures which contribute to many of the world's most acute environmental problems. Moreover, the process is circular, since it is poor people and poor countries which depend more than others on land and natural resources for survival and which are consequently more vulnerable to environmental deterioration.

It is important to realise that these problems do not arise from ignorance, let alone stupidity. There is, in most poor countries, a sophisticated awareness of the kind of agricultural practices that are sustainable. Particularly in India, China, Indonesia and also in many parts of Africa, there are, in peasant farming communities, traditions of terracing, crop rotation, natural fertilizers and animal husbandry that long pre-date the arrival of European technology. But poor countries often find themselves trapped in a downward spiral in which the pressures of poverty and rising population lead to sound practices being abandoned. There is also a generally wide awareness of the undesirability of excessive population growth: in a recent survey, while only 10 per cent of women in rural Ghana were practising contraception, 90 per cent expressed a clear preference for having no more children.[6] But partly because of the high levels of child mortality – and the generally low status of women – these wishes do not prevail over the seeming compulsions of economic need; they go unfulfilled, with disastrous consequences for development, and eventually for global living.

What makes all this so acutely critical is that all the signs point to the incidence of poverty growing in the Third World. For example, the number of people on inadequate diets, excluding China, rose from an estimated 650 million to 730 million in the 1970s and, since 1980, matters have turned from bad to worse. Among children under five alone, 160 million are reported to suffer protein energy malnutrition – and this includes two-thirds of all children in South Asia. In twenty-one out of thirty-five low income developing countries, the overall daily calorie supply per capita was lower in 1985 than in 1965.[7] Almost a half of 115 developing countries have experienced falling per capita staple food consumption this decade. In most parts of the developing world, there have been sharply reduced growth rates, falls in real per capita income, rising unemployment and cut-backs in educational and health provision as a result of austerity measures consequent upon economic crisis. Such poverty is the worst form of pollution.

Let no one dismiss this as rhetoric. The simple, and terrible, truth is that poverty and environment are inextricably linked in a chain of cause and effect. Problems of environment cannot be tackled in isolation from those national and international economic factors that perpetuate large-scale poverty. This is why those concerned with environment in Latin America and Africa see links with international economic problems such as oppressive debt servicing and depressed commodity export prices, which together force developing countries to over-exploit their natural resource base in order to maintain export earnings. Imagine, then, how utterly galling it is for them to find Western aid agencies and multilateral institutions (like the World Bank) preaching about the need for greater environmental concern in developing countries against a backdrop of grossly inadequate financial flows, which perpetuate the very underdevelopment which contributes to environmental neglect. Man has stood on the moon and looked at the earth's oneness; yet centuries of preoccupation with ourselves – in family, then tribe, then nation-state – still stand in the way of those holistic global perceptions and solutions that are essential to human survival.

Amidst this catalogue of negative trends it is easy to be defeatist. The message of the World Commission was, however, by no means a negative one; and, while, for example, the massive burning of fossil fuels continues to degrade the physical environment everywhere, some decidedly positive developments are now beginning to emerge. In Scandinavia, Canada and West Germany, the environment is at, or close to, the top of the political agenda; in Britain, Mrs Thatcher has made a strong and welcome intervention in this area; now, President Bush has appointed a strong conservationist to head his Environmental Protection Agency. It is, of course, somewhat easier to espouse and pursue environmentalism in rich countries where there are no great pressures of rising population and where resources can be diverted to environmental protection. But is it no less welcome for that. There has also been significant progress in recognising the need to curb sulphur emissions that cause acid rain, and in introducing more environmentally sensitive agricultural policies.

So far, most 'green' politics has been in, and for the benefit of, developed countries; but some developing country politicians have also taken up the challenge. Rajiv Gandhi launched a programme to turn back 5 million hectares of land every year into fuelwood and fodder plantations. Robert Mugabe has tackled the sensitive issue of population and family planning: within two years of his launching the programme, contraceptive practice in Zimbabwe rose from 14 per cent to 38 per cent.[8] Of comparable significance is the way President Gorbachev is trying to lead international opinion. When the World Commission visited Moscow in December 1986 the administration was reeling in the aftermath of Chernobyl, and was very much on the defensive. Soviet authorities now freely admit that, no less than in the West, great damage has been done to the Russian environment by insensitive industrial planning. The planning system has quite disastrously neglected environmental factors and has promoted, through irrational pricing, a wasteful use of energy and raw materials. The emergence of the Soviet Union as a major and constructive participant in global environmental discussions adds a whole new dimension to what is possible.

These stirrings of awareness at a national level are now beginning to create the basis for tackling those environmental problems that are truly international. The most striking success has been the international agreement, under UNEP auspices, to cut the production and use of CFCs. It has been estimated that the Montreal Accord, if fully implemented, could avoid two million future skin cancer deaths.[9] Even before the agreement came into effect, however, scientific research suggested that far more dramatic curbs will be necessary to prevent continuing increases in skin cancer due to ozone depletion. Still, the agreement was remarkable from several standpoints. It was the first time governments had acted together, not in response to a demonstrated calamity, but to the predictions and warnings of scientists. It imposed effective restraints on some of the world's most powerful multinational companies. And, at a time when the whole idea of multilateralism has been called into question, it showed that global cooperation to face global threats is, after all, possible.

The agreement is propitious, too, for dealing with the much bigger and more complex problems of global warming produced by greenhouse gases, notably carbon dioxide. Within the last few months, a major step forward has been taken with the formulation of a clear consensus among scientists of the scale of global warming that is likely. The Expert Group, which the Commonwealth has established in the light of the Vancouver summit discussions to look at the issue of climate change, estimates that, even with policies adopted now to reduce greenhouse gas emissions, there is a 90 per cent probability of mean global warming of at least 1 to 2 degrees by the year 2030 and continued warming after that.[10] Some estimates are of much higher figures. I deliberately present the estimates in a cautious way since there is a danger of environmentalists overstating their case and inviting the kind of ridicule which one leading columnist captured when he claimed to have read that 'if we continue to use underarm deodorants, drive motor cars and burn fossil fuels unchecked the consequent greenhouse effect may create temperatures at which lead melts and other metals become red hot'.[11] Even so, scientific consensus does suggest that the speed of warming – which is already being monitored – is historically unprecedented.

While the precise effects of this process on particular countries or even regions cannot yet be predicted with any certainty, among the more probable will be growing aridity in already semi-arid tropical areas – those parts of Africa, for example, that have recently experienced severe drought and famine. Another will be a greater incidence of extreme events, such as major hurricanes in tropical areas. The US Space Agency (NASA) has suggested that hurricane 'Gilbert' was a harbinger of a more powerful and disastrous generation of hurricanes engendered by global warming.[12] Even small climatic changes can have dramatic effects. It is believed, for example, that the lethal strain of mosquito which has killed tens of thousands of people in Madagascar in the last year, could have flourished due to slight warming.[13] There may be some beneficiaries if, for example, winter warming in Canada and the USSR permits agriculture in the more northerly, currently frozen, latitudes. But there is no basis for complacency that global warming could be a 'zero sum game'; adjustment will be necessary everywhere, with attendant costs.

Among these costs will be those of adjusting to higher sea levels. The Commonwealth Expert Group conservatively predicts, again with a very high level of probability, that the sea will rise at least 20 to 30 cm. (approximately 12 inches) by 2030 and possibly as much as three times that level with continuing increases for decades – perhaps for centuries – to come.[14] For areas at, or below, sea level – the big, highly populated deltas of Egypt, India and China; large areas of the United States, Britain and Holland; coastal atolls in the Indian and Pacific oceans; and the capital of my own country Guyana, which is built behind dykes – there is the prospect of widespread, perhaps catastrophic, flooding in years to come.[15]

Surveys of some of these areas conducted for the Commonwealth Group suggest brutal options. One is large-scale abandonment of land; conceivably, abandonment of whole countries. In the case of the Maldives, for example, the overwhelming majority of the 1200 islands in the chain are already barely above sea level. Who will house the displaced population of low-lying areas? Current attitudes to refugees and immigrants in most countries do not suggest that large population movements are feasible. Acceptance of an enhanced risk of large-scale drowning is clearly not an option. The 1970 cyclone in Bangladesh, when 300,000 were killed in one storm surge, and 1988's lesser though serious disaster, when hundreds more perished, are warnings of what could happen on an even larger scale.[16] Common humanity alone prevents us regarding this as an acceptable risk. A possible option is to build defences. But this is simply beyond the means of most poor countries. A single 4-kilometre barrier in Holland cost over $3bn; many countries would require much more. Is it not a global challenge to make our planet habitable for all its people? And is the challenge not at hand?

The costs of doing nothing to prevent climate change are simply unacceptable. Concern is now such that for the first time, following a little-noted meeting in Geneva in November 1988, governments have at least started to consider, under UN auspices, the issue of climatic change and global warming collectively. The problems in progressing from collective study to collective action are, however, immense. There is no obvious way of stopping some greenhouse gas accumulations. The Montreal Accord on CFCs should be of some indirect help, as would the slowing-down of deforestation. The core issue, however, is the carbon dioxide emitted by burning carbon-based fuel, especially coal. The main clean, renewable, source of energy – nuclear power – is, to say the least, problematic as well as costly and is likely to remain so, especially for countries with a limited technological base. Non-conventional sources of energy have been shown to be useful at the margin; but cannot substitute for traditional sources in a short time scale.

The approach of many environmentalists to this dilemma is to advocate a world of slower economic growth. While this may be superficially appealing to those already materially comfortable, it is both selfish and unwise. Given the extent and growth of mass poverty and the link between poverty and environmental stress, rapid economic growth in developing countries is essential; also developing countries can grow more rapidly in a buoyant world economy that bolsters trade opportunities, particularly commodity markets. The World Commission spoke of 5 per cent as representing a rough normative minimum economic growth for developing countries taken as a whole. At present, only a small number of countries, mainly in Asia, are reaching that level. Some parts of the Third World have experienced much lower economic growth in the last decade or so – notably Africa and Latin America – and have experienced declining per capita incomes, deepening poverty and the most extreme forms of environmental stress. Unless growth is revived there is no prospect for reversing these trends. And growth is necessary not only for developing countries; '*perestroika*' in the Soviet Union is quintessentially concerned with transforming stagnation into growth. And few seriously imagine that the major problems of the West, with large pockets of poverty and unemployment, and fraying public services and infrastructure, can be solved except in dynamic economies. The last decade has seen, in almost every part of the world, a combination of slower economic growth and accelerated environmental decline. The experience does not commend itself for the future.

So, as long as large-scale poverty and rapid population growth remain, 'no growth' is no solution. The Brundtland Commission made a major break with earlier environmental analysis – such as the Report of the Club of Rome in the early 1970s – by recognising this explicitly. It looked forward positively to 'a new era of growth'. But it stressed that growth must be qualitatively different from that experienced in the past; it must be growth that contributes to sustainable development: as we defined it, 'progress (in all countries) that meets the needs of the present without compromising the ability of future generations to meet their own needs'. This means, for example, not being mesmerised by GNP figures; in some countries these show impressive progress which is entirely illusory when we take into account the depletion of forests and other environmental assets, and the quality of life.

In most respects, growth and environmental sustainability reinforce each other naturally. This is the case in agriculture. There is now abundant evidence that the kind of protective CAP system operated by the EEC, and comparable arrangements in Japan and elsewhere, have both retarded economic efficiency and added greatly to the pressures for excessive, and chemical-intensive, production; while in developing countries, the artificial suppression of farm prices in the interest of urban consumers has retarded economic development and made environmentally prudent farming uneconomic.

Equally, there are some real conflicts of interest. Growth, even if carefully managed, is bound to result in a rapid growth of demand for some natural resources, particularly energy. Electricity generation multiplied eight times worldwide between 1950 and 1980. The electrification of villages; the replacement of wood burning by commercial power; industrialisation: these are integral parts of development. But technology provides at least the hope that it will be possible to counter this problem. Over the last decade, Western countries have achieved an annual growth of energy efficiency of 2 per cent and the potential undoubtedly exists for further major savings and on a worldwide basis.[17] It is possible to reconcile rapid growth with frugal energy and material use. But it will not be easy; it requires a willingness by consumers to pay high prices that fully reflect their environmental as well as narrowly economic costs; and governments to tax depletable resources and to finance major long-term research, difficult enough in secure democracies, let alone in countries where government is fragile and the rise in petrol and kerosene prices can precipitate riots and even *coup d'etats*. Fresh water supplies, of which there are reported scarcities in eighty countries, now provide an even more acute constraint on expansion, particularly of agriculture.

These problems will present painful dilemmas to which technology will not always provide solutions. It is, however, possible to say with reasonable confidence that those countries which are able to surmount the external and internal impediments to growth will find it easier to deal with the dilemmas. Korea, for example, is one of the very few developing countries expanding its forest acreage; Singapore has become the very antithesis of urban squalor; among the less publicised developmental success stories, Cyprus has been a leader in energy conservation and solar power, and Mauritius is pioneering a comprehensive approach to economic and environmental planning.

While there are pieces of evidence and experience which suggest that brisk economic growth and respect for enviromental values can be reconciled, we would be fooling ourselves if we imagine that 'sustainable development' is as yet well established as a working principle. To achieve it globally will require a transformation of attitudes in some fundamental respects. It requires, first of all, a long-term perspective; a recognition that we all have an obligation to future generations as well as to ourselves. This is difficult to realise where key policy makers in the rich world are geared to the daily stock exchange index and monthly balance of payment figures, and in the poor world to watching the level of foodstocks in granaries and foreign exchange reserves at the Central Bank. Even when priorities shift from the immediate to issues of long-term investment, investment criteria in both public and private sectors invariably discount future costs and benefits, so that the very long-term is always effectively ignored. In almost every country policies are being formulated whose long-term environmental implications – for climate change, sea level rise, species extinction and nuclear waste disposal – are not part of the calculus of decision-making. This has to change.

A second imperative which deserves to be stressed is the need to see environmental problems in interdisciplinary terms, not in terms of narrow specialisations. The world is replete with projects that made excellent engineering sense but were economically disastrous; or were economically sound but environmentally catastrophic. The current work on greeenhouse gases and global warming and their effects requires the combined skills of physicists, meteorologists, oceanographers, biologists, geographers, economists, lawyers, engineers and students of international relations, among others. To deal with such problems satisfactorily is a challenge to both statesmen and thinkers. There are pressures – of institutions, of culture, of fashion – to work in national and disciplinary compartments. These must be resisted.

Finally – and this is a factor I particularly stress as the head of an international organisation – a large and growing number of environmental issues are cross-border problems which simply cannot be solved nationally. Norwegian lakes and trees have been polluted by power stations in England and East Germany; the Chernobyl fallout affected farmers as far afield as Wales and Scotland; tree-cutting in Nepal has led to flooding in Bangladesh, and in Ethiopia it has caused water supply problems in Sudan and Egypt; CFC emissions in the northern hemisphere could be the cause of skin cancer in Australia, Chile and Argentina. Unless there is a regional or global framework for handling such issues we will see some of them escalating dangerously, in some cases to conflict, as may already have occurred as a consequence of large-scale environmental refugee movements in the Horn of Africa. It is not at all implausible to hypothesise that if environment and development problems become even more serious (with, for example, the large-scale involuntary migrations many believe will follow climatic change) some of the most serious consequences could be in the field of security.

Some international problems – those concerned with the global commons, the oceans and the ocean bed, the atmosphere, Antarctica, which no one owns – present particular difficulties. Unless access is regulated in some way, rising demand will result in over-use. For the global commons this means internationally-agreed controls. In some respects, such as deep-sea fishing, dumping of waste, fisheries agreements and Antarctica, there have been embryonic forms of multilateral control. But they tread a very delicate dividing line between the competing claims of conservation, private business and governments, all with different interests. To handle these common problems requires strong multilateral institutions and respect for international law. That means a change in habits by some of the major powers.

An effective Law of the Sea to manage the ocean bed has been frustrated for the last decade by the refusal of the US to conform. Russia and Japan have often shown a cavalier disregard for the need to observe fishing agreements. More seriously, there is insufficient attention to the position of poorer countries which are trying to develop in a world from which much of the environmental capital has already been drawn and effective control of much of the remainder lies elsewhere. For example, the treaty governing Antarctica, in some respects an admirably enlightened and conservationist arrangement, which has kept the world's last true wilderness free of both weapons and developers, is currently faced with the issue of whether or not to allow mineral and oil exploration. The decision will be made only by the eighteen treaty countries; the decision-making process will, for example, have no African representation, except for South Africa. Regulation of all the world's global commons, which are truly part of the common heritage of mankind, face similar problems of inequity and unrepresentative control.

An enduring message of the Brundtland Report was that questions of the global environment cannot be separated from the political, economic and moral issues posed by a world in which there is great wealth and also great poverty; with states trying to coexist that range from the superpowers to vulnerable microstates; and with still only tenuous legal and institutional arrangements preserving international order. Underlining this message of a common future was the unspoken premise that we must think of our planet not only as a world of many states but also as the state of our one world; that we must be ready to nurture tomorrow's concepts of global governance, not have them stifled at birth by yesterday's notions of national sovereignty; that our common future may not be secured save by the reach of enforceable law across environmentally invisible frontiers. I hope that in dealing with the technical issues, we will not lose sight of this wider, and necessary, dimension of saving our endangered earth.

2 Threatened Islands, Threatened Earth; Early Professional Science and the Historical Origins of Global Environmental Concerns

Richard Grove

In recent years we have seen an explosion of popular and governmental interest in environmental problems. The world is widely seen to be in the throes of an environmental crisis, in which an artificially-induced 'greenhouse effect' hangs over humanity like a climatic Sword of Damocles. As a result, environmental matters have become a critical part of the political agenda in almost every country. Increasingly, too, the prescriptions of environmentalists are receiving popular acclaim and support of a kind that, before now, was heard only from a minority. Ideas about conservation and sustainable development, in particular, have become highly politicised.

It is clearly right that the environmental future of the earth should be a matter of popular preoccupation. It has, however, helped to bring about a widespread belief that environmental concerns are an entirely new matter, and that conservationist attempts to intervene in human despoliation of the earth are part of a new and revolutionary programme. However, while the degree of popular interest in global environmental degradation may be something new, the histories of environmental concern and conservation are certainly not. On the contrary, the origins and early history of contemporary Western environmental concern and concomitant attempts at conservationist intervention lie far back in time. To take one particular instance, the fear of widespread artificially-induced climate change, widely thought to be of recent origin, actually has roots in the writings of Theophrastus and others in classical Greece,[1] and formed the basis for the first forest conservationist policies of the British colonial states. As early as the mid-eighteenth century, scientists were able to manipulate state policy by their capacity to play on fears of environmental cataclysm, just as they are today. By 1850 the problem of tropical deforestation was already being conceived of as a problem existing on a continental scale, and as a phenomenon demanding urgent and concerted state intervention. Now that scientists and environmentalists once again have the upper hand in state environmental policy, we may do well to recall the story of their first, relatively short-lived, periods of power.

Early scientific critiques of 'development' or 'improvement' were, in fact, well developed by the early nineteenth century. The fact that such critiques emerged under the conditions of colonial rule in the tropics is not altogether surprising. The kind of homogenising, capital-intensive transformation of people, trade, economy and environment with which we are familiar today can be traced back to the beginnings of European colonial expansion, as the agents of new European capital and urban markets sought to extend their areas of operation and sources of raw material supply. It is clearly important, therefore, to try to understand current environmental concerns in the light of a long historical perspective on the diffusion of capital-intensive Western (and occasionally indigenous) economic forces. Similarly, the evolution of a reasoned awareness of the wholesale vulnerability of earth to man and the idea of 'conservation', particularly as practised by the state, has been closely informed by the evolution of a global environmental knowledge.

Early environmental concerns, and critiques of the impact of Western economic forces on tropical environments in particular, emerged as a corollary of, and in some sense as a contradiction to, the history of the mental and material colonisation of the world by Europeans. To date, most attempts to understand the emergence of purposive and conservationist responses to the destructive impact of man on nature have been largely confined to localised European and North American contexts. Early environmentalism has generally been interpreted as a local response to the conditions of Western industrialisation, while conservation has been seen as deriving from a specifically North American setting.[2] Moreover, the fact that Anglo-Americans such as George Perkins Marsh,[3] Henry David Thoreau and Theodore Roosevelt have been elevated to a pantheon of conservationist prophets has discouraged the proper investigation of even their earlier European counterparts, let alone those of elsewhere.[4]

All this has meant that the real and far more complex antecedents of contemporary conservationist attitudes and policies have quite simply been overlooked in the absence of any attempt to deal with the history of environmental concern on a truly global basis. In particular, and largely for quite understandable ideological reasons, very little account has ever been taken of the significance of the colonial experience on the formation of Western environmental attitudes and critiques. Furthermore, the crucially pervasive and creative impact of the tropical and colonial experience on natural science, and on the Western and scientific mind after the fifteenth century, has been almost entirely ignored by those environmental historians and geographers who have sought to disentangle the history of environmentalism and changing attitudes to nature.[5] Added to this, the historically decisive diffusion of indigenous, and particularly Indian, environmental philosophy and knowledge into Western thought after the late eighteenth century has been largely dismissed. Instead, it has simply been assumed that European and colonial attempts to respond to tropical environmental change derived exclusively from metropolitan and northern models and attitudes.

In fact, the converse was true. The available evidence shows that the seeds of modern conservationism developed as an integral part of the European encounter with the tropics. As colonial expansion proceeded, the environmental experiences of Europeans and indigenous peoples living at the colonial periphery played a steadily more dominant and dynamic part in the development of new European evaluations of nature and in the growing awareness of the destructive impact of European economic activity on the peoples and environments of the newly 'discovered' and colonised lands.[6] After the fifteenth century the emerging global framework of trade and travel provided the conditions for a process by which indigenous European notions about 'nature' were gradually transformed, or even submerged, by a plethora of information, impressions and inspiration from the wider world.[7] In this way, the commercial and utilitarian purposes of European expansion produced a situation in which the tropical environment was increasingly utilised as the symbolic location for the idealised landscapes and aspirations of the western imagination. William Shakespeare's play *The Tempest* and Andrew Marvell's poem 'Bermoothes' stand as pioneering literary exemplars of this cultural trend.[8] These aspirations now became global in their scope and reach and increasingly exerted an influence on the way in which newly colonised lands and peoples were organised and appropriated.

The notion that the Garden and Rivers of Eden might be discovered somewhere in the East was already a very ancient one in European myth.[9] Early Renaissance conceptions of Eden or Paradise, which often took concrete shape in the form of the early systematic botanic gardens, were derived from Persian and Mughal notions of 'Paradis' (or 'a garden') that had originated in Persia, Northern India and the Middle East.[10] The developing scope of European expansion during the Renaissance offered the opportunity for this search for Eden and the dyadic 'other' to be realised and expanded as a great project and partner of the other more obviously economic projects of early colonialism.[11] Later, the search for Eden provided much of the imaginative basis for early Romanticism, whose visual symbols were frequently located in the tropics, and for late eighteenth-century Orientalism, for which the Edenic search was as essential precursor.[12] Northern India, and the Ganges valley had, after all, long been considered as a much-favoured candidate for the location of the Garden of Eden.[13] Some researchers have suggested that Judaeo-Christian attitudes to the environment have been inherently destructive. Such claims are highly debatable.[14] In fact, the rapid ecological changes caused by the inherently transforming potential of capital in the context of European expansion have tended to be confused with the consequences of doctrinally specific attitudes to the environment.[15] There were, however, clear links between religious change during the sixteenth century and the emergence of a more sympathetic environmental psychology. Above all, the advent of Protestantism in Europe seems to have lent a further impetus to the Edenic search, as a knowledge of the natural world began to be seen as a respectable path to seeking knowledge of God.

During the fifteenth century the task of locating Eden and re-evaluating nature had already begun to be served by the appropriation of the newly-discovered and colonised tropical islands as Paradises. Dante's citing of an 'Earthly Paradise' in a 'southern ocean' is a case in point.[16] This role was reinforced by the establishment of the earliest colonial botanic gardens on these islands and on one mainland 'Eden', the Cape of Good Hope. These imaginative projections were not, however, easily confined.[17] Conceptually, they soon expanded beyond the botanic garden to encompass large tropical islands. Subsequently, the colonialist encounter in India, Africa and the Americas with large 'wild' landscapes apparently little altered by man, along with their huge variety of plants (no longer confinable, as they had been, to one botanic garden), meant that the whole tropical world became vulnerable to colonisation by an ever-expanding and ambitous imaginative symbolism. Frequently, such notions were closely allied to stereotyping of the luckless indigenous population as 'noble savages'.[18] Ultimately, then, the area of the new and far more complex European 'Eden' of the late eighteenth and early nineteenth centuries knew no real bounds. Even Australia and Antarctica, in recent years, have not been immune to being termed as 'Edens'. The imaginative hegemony implied by new valuations of nature, which had themselves been stimulated by the encounter with the colonial periphery, had enormous implications for the way in which the real, economic, impact of the coloniser on the natural environment was assessed by the new ecological critics of colonial rule. This assessment rarely included, however, much appreciation of any local or indigenous understanding of the environment, thus creating a negative tendency which has persisted until the present day.

Paradoxically, the full flowering of what one might term the Edenic island 'discourse' during the mid-seventeenth century closely coincided with the realisation that the economic demands of colonial rule on previously uninhabited oceanic island colonies threatened their imminent and comprehensive despoliation. Extensive descriptions exist of the damaging ecological effects of deforestation and European plantation agriculture on the Canary Islands and Madeira (the 'wooded isle') after about 1300, and in the West Indies after 1560.[19] In the Canary Islands complex irrigation systems formed part of an early technical response to the desiccation that followed despoliation. In the West Indies, particularly on Barbados[20] and Jamaica, local attempts were made to try to prevent excessive soil erosion in the wake of clearance for plantations.[21] Some of the worst consequences of early colonial deforestation are well documented on the island colonies of St Helena and Mauritius. It was on these islands that a coherent and wide-ranging critique of environmental degradation first emerged in the context of colonial annexation and economic development.

It is certainly true that anxieties about soil erosion and deforestation had emerged at earlier periods, in the literature of classical Greece, imperial Rome and Mauryan India, and then in a sporadic and unconnected fashion in the annals of the early Spanish and Portuguese colonial empires. For example, in 450 BC Artaxerxes I attempted to restrict the cutting of the cedars of Lebanon. A little later the Mauryan Kings of northern India adopted a highly organised system of forest reserves and elephant protection. Similarly, indigenous strategies for environmental management on a small scale, involving a considerable understanding of environmental processes, had existed in many parts of the world since time immemorial. But it was not until the mid-seventeenth century that a coherent and relatively organised awareness of the ecological impact of the demands of emergent capitalism and colonial rule started to emerge and to grow into a fully-fledged theory about the limitability of the natural resources of the earth and the need for conservation.

This new sensitivity developed, ironically, as a product of the very specific, and ecologically destructive, conditions of the commercial expansion of the Dutch and English East India Companies and, a little later, of the Compagnie des Indes. The conservationist ideology to which it gave rise was based both upon a new kind of evaluation of tropical nature and upon the highly empirical and geographically circumscribed observations of environmental processes which the experience of tropical island environments had made possible. It has been assumed by some historians that the colonial experience was not only highly destructive in environmental terms but that its very destructiveness had its roots in ideologically 'imperialist' attitudes towards the environment.[22] On the face of it this does not seem an extraordinary thesis to advance, particularly as the evidence would seem to indicate that the penetration of Western economic forces which colonial annexation facilitated did indeed promote a rapid ecological transformation in many parts of the world. This was especially true in the late nineteenth century in Southern Africa where a particularly exploitative agricultural and hunting ethos at first prevailed.[23]

On closer inspection, however, the hypothesis of a purely destructive 'environmental imperialism' does not appear to stand up at all well. Indeed, it has become increasingly clear that there is a need to question the more monolithic 'imperialist' hypotheses which seem to have arisen out of a misunderstanding of the essentially heterogenous and ambivalent nature of the workings of the early colonial state. Many scholars have remained unaware of the extent to which many colonial states were peculiarly open, during the first half of the nineteenth century, to the social leverage and often radical agendas of the contemporary scientific lobby, at a time of great uncertainty about the role, and the long-term security, of colonial rule. Furthermore, while the colonial enterprise undoubtedly promoted large-scale ecological change, it also helped to create a context that was conducive to rigorous analytical thinking about the actual processes of ecological change as well as about the potential for new forms of land-control. Ironically, too, the colonial state in its pioneering conservationist role provided a forum for controls on the unhindered operations of capital for short-term gain which, it might be argued, brought about a contradiction to what is normally supposed to have made up the common currency of imperial expansion. Ultimately, the long-term security of the state, which any ecological crisis threatened to undermine, counted for far more than the interests of private capital bent on ecologically destructive transformation.[24] Indeed, the absolutist nature of colonial rule facilitated interventionist forms of land management that, at the time, would have been very difficult to impose in Europe. Colonial expansion also promoted the rapid diffusion of new scientific ideas between colonies, and between metropole and colony, over a large area of the world.[25]

The continuity and the survival of the kind of critique of the ecological impact of colonial 'development' which had developed by the early eighteenth century was facilitated by, and indeed dependent on, the presence of a coterie of committed professional scientists and environmental commentators. These men, almost all of whom were medical surgeons and custodians of the early colonial botanic gardens, were already an essential part of the administrative and hierarchical machinery of the new trading companies. As East India Company investment in trade expanded into an investment in territorial acquisition, the members of the medical and botanical branches grew steadily in number. In India, for example, in 1838, there were over 800 surgeons employed at one time in different parts of the Company's possessions.[26] As time passed, more and more complex administrative and technical demands were made upon these highly-educated and often independent-thinking colonial employees.[27] During the early eighteenth century the urgent need to understand unfamiliar florae, faunae and geologies, both for commercial purposes and to counter environmental health risks, propelled many erstwhile physicians and surgeons into consulting positions and employment with the trading companies as fully-fledged professional and state scientists, long before such a phenomenon existed in Europe. By the end of the eighteenth century their new environmental theories, along with an ever-growing flood of information about the natural history and ethnology of the newly colonised lands, were quickly diffused through the meetings and publications of a whole set of 'academies' and scientific societies based throughout the colonial world.

The first of these had developed in the early island colonies, particularly on Mauritius, where the Baconian organising traditions of the metropolitan institutions of Colbert found an entirely new purpose.[28] This was no accident. In many respects the isolated oceanic island, like the frail ships on the great scientific circumnavigations of the seventeenth and eighteenth centuries, directly stimulated the emergence of a detached self-consciousness and a critical view of European origins and behaviour, of the kind dramatically prefigured by Daniel Defoe in *Robinson Crusoe*. Thus the island easily became, in practical environmental as well as mental terms, an easily-conceived allegory of a whole world. Contemporary observations of their ecological demise were easily converted into premonitions of environmental destruction on a more global scale. Thus, alongside the emergence of professional natural science, the importance of the island as a mental symbol continued to constitute a critical stimulant to the development of concepts of environmental protection, as well as of ethnological identity. Half a century before an acquaintance with the Falkland[29] and Galapagos islands provided Charles Darwin with the data he required to construct a theory of evolution, the isolated and peculiar floras of St Helena and Mauritius had already sown the seeds for a concept of rarity and a fear of extinctions that, by the 1770s, were already well developed in the minds of French and British colonial botanists. Furthermore, the scientific odysseys of Bougainville and Cook served to reinforce the significance of tropical islands, Otaheite and Mauritius in particular, as symbolic and practical locations of the social and physical Utopias beloved of the early Romantic reaction to the Enlightenment.[30] The environments of tropical islands thus became even more highly prized, so that it may come as no surprise to discover that it was upon one of them, Mauritius, that the early environmental debate came to a head. Under the influence of zealous French anti-capitalist Physiocrat reformers and their successors between 1768 and 1810, this island became the location for some of the earliest comprehensive experiments in systematic forest conservation, pollution control and fisheries protection. These initiatives were carried out by scientists who, characteristically, were both followers of Jean-Jacques Rousseau and adherents of the kind of rigorous empiricism associated with mid-eighteenth century French Enlightenment botany. Their forest conservation measures were based on a highly developed awareness of the potentially global impact of modern economic activity, on a fear of the climatic consequence of deforestation and, not least, on a fear of species extinctions. As a consequence the 'Romantic' scientists of Mauritius, and above all Pierre Poivre, Philibert Commerson and Bernadin de St Pierre can, from hindsight, be seen as being the pioneers of modern environmentalism.[31] All of them saw a responsible stewardship of the environment as a priority of aesthetic and moral economy as well as a matter of economic necessity.

After the annexation of Mauritius by the British in 1810, such nascent environmental notions and conservationist prescriptions were transferred to St Helena and eventually to India itself by a now severely worried East India Company. There is some irony in this as the

Company had in earlier times simply ignored the cries for help of the St Helena governors, isolated as they were in the midst of fast-moving ecological crisis. The apparently successful results of tree-planting and other environmental protection policies on Mauritius and St Helena eventually provided much of the justification and many of the practical models for the early forest planting and conservancy systems which developed in India and elsewhere after the early 1830s. It appears that, until then, the emergence of concerns about the effects of environmental change had been delayed by the sheer scale of the sub-continent, which served effectively to conceal the effects of soil erosion and deforestation.

The Edenic, Romantic and Physiocratic roots of environmentalism in Mauritius and India were strongly reinforced after 1820 by the writings of Alexander von Humboldt. Pierre Poivre, on Mauritius, had already been persuaded of the value of tree-planting and protection by his observations of Indian and Chinese forestry and horticultural methods and his knowledge of Dutch botanic gardening techniques derived, circuitously, from the Mughal emperors. Humboldt's environmental writings, however, were guided by Indian thinking in a far more profound way. Much influenced by the seminal Orientalist writings of Johann Herder, as well as by those of his own brother, Wilhelm, Alexander von Humboldt strove in successive books to promulgate a new ecological concept of the relations between man and the natural world which was drawn almost entirely from the characteristically holist and unitary thinking of Hindu philosophers. His theoretical subordination of man to other forces in the cosmos formed the basis for a universalist and scientifically-reasoned interpretation of the ecological threat posed by the unrestrained activities of man.

This interpretation became particularly influential among the Scottish scientists employed by the East India Company. These men, mainly medical surgeons trained in the rigorous French-derived Enlightenment traditions of Edinburgh, Glasgow and Aberdeen Universities, were especially receptive to a mode of thinking which related the multiple factors of deforestation, water supply, famine, climate and disease in a clear and coherent fashion.[32] Several of them, in particular Alexander Gibson, Edward Balfour and Hugh Cleghorn, became enthusiastic proselytisers of a conservationist message which proved both highly alarming to the East India Company, and highly effective in providing the ideological basis for the pioneering of a forest conservancy system in India on a hitherto unequalled scale. The environmental views of the East Indian Company surgeons were most effectively summed up in a report published in 1852 entitled 'Report of a committee appointed by the British Association to consider the probable effects in an economic and physical point of view of the destruction of the tropical forests'. This warned that a failure to set up an effective forest protection system would result in ecological and social disaster. Its authors were able to point to the massive deforestation and soil erosion which had occurred on the Malabar coast, with the resulting silting-up of commercially important harbours, as early evidence of what might happen in the absence of a state conservation programme. The report took a global approach, drawing on evidence and scientific papers from all over the world, and did not confine its analysis to India. Later, the forest conservation system set up in India (based in part on Mauritius experience) provided the model for most of the systems of colonial state conservation which developed in South-East Asia, Australasia and Africa and much later, in North America. It should be noted, however, that the first warnings of the dangers of deforestation in India that were communicated to the colonial authorities had originated with indigenous Indian rulers as much as with the early colonial scientists.[33]

To summarise, the ideological and scientific content of early colonial conservationism as it had developed under early British and French colonial rule by the 1850s amounted to a highly heterogenous mixture of indigenous, Romantic, Orientalist and other elements. The thinking of the scientific pioneers of early conservationism was often contradictory and confused. Many of their prescriptions were constrained by the needs of the colonial state, even though the state at first resisted the notion of conservation. In the second half of the nineteenth century, too, forest conservation and associated forced resettlement methods were frequently the cause of a fierce oppression of indigenous peoples and became a highly convenient form of social control.[34] Indeed, resistance to colonial conservation structures became a central element in the formation of many early anti-imperialist nationalist movements.[35] But despite the overarching priorities and distortions of colonialism, the early colonial conservationists nevertheless remain entitled to occupy a very important historical niche. This is above all because they were able to foresee, with remarkable precision, the apparently unmanageable environmental problems of today. Their antecedents, motivations and agendas thus demand our close attention.

If there is a single historical lesson to be drawn from the early history of conservation under the East India Company and the *ancien règime* on Mauritius, it is that states can only be persuaded to act to prevent environmental degradation when their economic interests are shown to be directly threatened. Time and again, from the mid-eighteenth century onwards, scientists discovered that the threat of artificially-induced climatic change, with all it implied, was one of the few really effective instruments that could be employed in persuading governments of the seriousness of environmental crisis. The argument that rapid deforestation might cause severe rainfall decline, decline in runoff and, eventually, famine, was one that was quickly grasped by the East India Company, fearful as it always was of agrarian economic failure and social unrest.

Unfortunately, it often required an initial famine to lend credibility to scientists in the eyes of government, and to provide the required impetus for the state to intervene with environmental protection measures. In India, for example, serious drought periods in 1835–9, the early 1860s, and 1877–8 were all rapidly followed by the initiation of renewal of state programmes designed to strengthen forest protection, often with the specific aim of preventing subsequent droughts. Of course, such legislation had the convenient by-product of increasing state control over land and timber supplies (often at the expense of common rights). The central motive, however, was related to an underlying fear of climatic change. Likewise, the early pioneer of state conservation in the Cape Colony, John Crombie Brown, was only able to secure government agreement to new forest conservation and grassland-burning prevention measures after the disastrous drought of 1861–1863 had wrought havoc on settler agriculture throughout the colony.[36]

In fact, the Southern African drought of 1862 encouraged the development of a whole school of 'desiccationist' theory, closely related to a contemporary Indian counterpart, which was convinced that most of the semi-arid tropics was undergoing long-term aridification as part of a process aided by colonial deforestation. Theories of widespread climatic change acquired further credibility when a paper was read at the Royal Geographical Society in London in March, 1865 on 'the progressing desiccation of the basin of the Orange river in Southern Africa' by James Fox Wilson, a naturalist and traveller. He believed that the Orange river was 'gradually becoming deprived of moisture' and that 'the Kalahari desert was gaining in extent'.[37] Wilson believed that the desiccation was due to 'the reckless burning of timber and the burning of pasture over many generations by natives'. David Livingstone, present at his lecture at the RGS, disagreed strongly with Wilson's analysis. The cause of rainfall decline, Livingstone asserted, was a natural phenomenon due to geo-physical phenomena. Another speaker, a Mr Galton, believed that the introduction of cheap axes into Africa by Europeans had promoted excessive deforestation and consequent drought. Yet another member of Wilson's audience, Colonel George Balfour, of the Indian Army, struck a more caustic note. Rainfall decline in India, he believed, was caused principally by the deforestation activities of the whole community, including European plantation owners. Counter-measures were necessary. He had been informed that morning, he said, that in the West Indies the Government of Trinidad had passed a law prohibiting the cutting down of trees near the capital, in order to ensure a supply of rain. Balfour was quick to point out on this, as on other occasions, that in pre-colonial times it had been the practice of Indians to sink wells and 'plant topes of trees' to encourage water retention. In another Royal Geographical Society debate, in 1866, Balfour pointed out, too, that 'in the Mauritius the Government had passed laws to prevent the cutting down of trees, and the result has been to secure an abundant supply of rainfall'.

The debate about climatic change had thus become international in reference and relevance by the mid-1860s. It was reinforced by more detailed research that raised the possibility that the very constitution of the atmosphere might be changing. Such views, the origins of the current 'greenhouse' debate, had found an early advocacy in the writings of J. Spotswood Wilson, who presented a paper in 1858 to the British Association for the Advancement of Science on 'the general and gradual desiccation of the earth and atmosphere'.[38] This paper probably helped to influence the ideas of the debaters at the Royal Geographical Society in 1865–1866. The upheaval of the land, 'destruction of forests and waste by irrigation' were not sufficient to explain the available facts on climate change, the author stated. The cause, he believed, lay in the changing proportions of oxygen and carbonic acid in the atmosphere.[39] Their respective ratios, he believed were connected with the relative rates of their production and absorption by the 'animal and vegetable kingdom'.[40] The author of this precocious paper concluded with a dismal set of remarks. Changes in 'the atmosphere and water' were

> in the usual course of geological changes, slowly approaching a state in which it will be impossible for man to continue as an inhabitant... as inferior races preceded man and enjoyed existence before the earth had arrived at a state suitable to his constitution, it is more probable that others will succeed him when the conditions necessary for his existence have passed away.

The raising, as early as 1858, of such a spectre of human extinction as a consequence of climatic change was clearly a shocking psychological development at the time. It was consistent, however, with fears that had been developing among the emergent world scientific community for a considerable period. The concept of species rarity and the possibility of extinction had, in fact, existed since the mid-seventeenth century as the scope of Western biological knowledge started to embrace the whole tropical world. The extinction of the auroch in 1627 in Poland and the dodo by 1670 in Mauritius had made a considerable impact. Already, by 1680 in Poland, large areas of forest had been set aside, as at Belovezh, where hunting of all kinds was prohibited for the whole community. The contemporary survival of the wisent, or European bison, is attributable to this isolated and pioneering effort. The mechanisms of extinctions, and the contribution of man to the process, first began to be clearly grasped on the tropical island colonies. On St Helena, attempts were made as early as 1713 to protect the St Helena Redwood in the knowledge of its apparently imminent demise.[41] Eighty years later William Burchell, the Government Botanist of St Helena from 1805 to 1810, and one of the first scientists employed by the East India Company, had acquired a systematic knowledge of the endemic plants of the island and had explicitly recognised the probability of species extinctions.[42] In part, this recognition was based on an exceptional knowledge of world plant species. Burchell was also, however, keenly aware of the rapid and continuing deforestation which was taking place on the island despite attempts made to legislate against it as early as 1709. A knowledge of the threatened and endemic character of the St Helena flora can be found in the writings of both William Roxburgh, the renowned Superintendent of the Calcutta Botanic Garden, and Charles Darwin himself.

The idiosyncratic and exceptional efforts made by Charles Waterton to make his estate into a nature reserve constitute interesting precedents and indicate a growing awareness of the destructive potential of man that had developed by the 1830s in Britain. However his actions were exceptional, even eccentric. One of the interesting features of Charles Waterton's initiatives is that he had travelled very extensively in South America, shortly after Humboldt, and there is little doubt that his enthusiasm for natural history and for the fate of species in Britain was closely associated with his enthusiasm for tropical wildlife and that of tropical forests in particular.[43]

The publication of the *Principles of Geology* by Charles Lyell in 1834, a book which questioned the fixity of species and laid the basis for the modern understanding of geological change, gave a firm footing to the confused awareness of the phenomenon of extinctions that had already developed among East India Company scientists such as Burchell on St Helena and William Roxburgh in Calcutta.[44] The question-mark which Lyell had placed over Genesis-based concepts of geological processes served to stimulate a revolution in earlier notions about the speed of environmental processes and, paradoxically, to emphasise the apparent helplessness of man in the face of environmental change.[45] By the early 1840s Ernst Dieffenbach, in his work on the fauna of New Zealand and the Chatham Islands, and then in his studies of Mauritius, had become more acutely aware of the potential for further rapid extinctions as European economic activity spread.[46] Hugh Strickland, first made aware of the extinction problem by his work on the palaeontology of the dodo and other extinct birds of the Mascarenes, actually suggested that the entire colony of New Zealand should be made a nature reserve to save its remaining indigenous fauna.[47] This was at a time when the value of rare island faunas was being recognised in the formulation of theories about species origins, not only by Darwin but also by Strickland, Hooker and Deiffenbach.[48] The publication of *The Origin of Species* by Charles Darwin in 1859, with its central emphasis on the place of extinction in the dynamics of natural selection, served only to sharpen the psychological dilemma of colonial scientists, many of whom were already aware of the part played by man in hastening species extinctions. A central part of the response to the existential havoc created by the *Origin* consisted, ultimately, in the acceleration of attempts to promote state conservation legislation. The publication of Darwin's theories had the effect of sharpening the dilemma of scientists who were already aware of the rapidly increasing contribution being made by man to the extinction process.

In fact this knowledge, crystallised by Darwin's work, underlay many of the attempts made by early environmentalists to secure state controls over tropical deforestation. Dr Hugh Cleghorn, for example, the first Inspector-General of the Madras Forest Department (set up in 1856), stated explicitly in several publications that uncontrolled deforestation would both cause the loss of potentially valuable species and prevent botanists from assembling the evidence for the evolution of species. He appears to have considered, however, that such arguments might not weigh convincingly with government, and chose to stress instead the more obvious economic hazards of climatic change and resource depletion.[49] The publication of *The Origin of Species*, however, went some way to making species protection a more valid concept in the eyes of government. Indeed, the decade 1860–70 produced a flurry of early attempts to legislate the protection of threatened species. Once more, the innovating context was an island colony, Tasmania, where a comprehensive body of laws, designed mainly to protect the indigenous birds, was introduced in 1860 after vigorous lobbying by J. Morton Allport, an amateur naturalist. Other territories rapidly followed suit. For example, the colonial legislatures of Natal, and Victoria (Australia), had both introduced laws to protect selected animal and bird species by 1865. Somewhat belatedly, the United Kingdom followed suit, introducing its first bird protection legislation in 1868. Significantly,the architect of the first British measures was Professor Alfred Newton, a frequent correspondent with Morton Allport in Tasmania, and the first prominent scientist to recognise the validity of Darwin's theory of natural selection. The timing of such early species protection measures, all closely connected with the climate of opinion stimulated by the seminal works of Lyell and Darwin, offered a symbolic as well as a practical opportunity to try to reassert control over a process of environmental degradation that was now understood as global in scope.

By the early 1860s, therefore, long-established anxieties about artificially-induced climatic change and species extinctions had reached a climax. The penetration of Western-style economic development, spread initially through colonial expansion, was increasingly seen by more perceptive scientists as threatening the survival of man himself. The subsequent evolution of the awareness of a global environmental threat has, to date, consisted almost entirely of a reiteration of a set of ideas that had reached full maturity over a century ago. It is to be regretted that it has taken so long for the warnings of early scientists to be taken seriously by advocates of unrestricted 'development'. Indeed, one hopes that it is not too late.

Part Two: Threats to the Environment

3 The Changing Climate and Problems of Prediction[1]

Stephen H. Schneider

The Earth's climate changes. It is vastly different now from what it was 100 million years ago, when dinosaurs dominated the planet and tropical plants thrived at high latitudes. It is different from what it was even 18,000 years ago, when ice sheets covered much more of the Northern Hemisphere. In the future it will surely continue to evolve. In part the evolution will continue to be driven by natural causes, such as slow changes in the Earth's orbit over many thousands of years. But future climatic change, unlike that of the past, will probably have another source as well: human activities. We may already be feeling the effects of having polluted the atmosphere with gases such as carbon dioxide, which play a major role in determining the earth's climate.

Carbon dioxide, and other so-called greenhouse gases, affect the temperature of the Earth by trapping heat near the surface. They are relatively transparent to visible sunlight, but absorb the long-wavelength, infrared radiation emitted by the Earth, retaining the heat. It is broadly accepted that when the carbon dioxide concentration rises, more radiation is absorbed, and the temperature at the Earth's surface will rise. That is the greenhouse effect, and its existence is not questioned.

A link between climatic change and fluctuations in greenhouse gases seems to be born out from evidence taken from the past. Laboratory analysis of gases trapped in cores of ice found in Antarctica, dating over a span of 160,000 years, has shown that carbon dioxide and methane levels varied in step with changes in the average local temperature. During the current interglacial period (the last 10,000 years), the ice cores have recorded a local temperature about 10 degrees centrigrade warmer during the height of the ice ages, and at the same time there has been about 25 per cent more carbon dioxide and about 100 per cent more methane in the atmosphere. It is not clear whether the greenhouse gas variations caused the climatic changes, or vice versa. It is possible that the ice ages were affected by other factors such as changes in the earth's orbital parameters, and that biological changes and ocean circulation shifts in turn affected atmospheric gas content, amplifying the climatic swings.

The last 100 years has seen a further 25 per cent increase of carbon dioxide and another doubling in the concentration of methane. There has also been about a half a degree centigrade of 'real' warming over this period, with the 1980s appearing to be the warmest decade. It is tempting, given such data, to accept this as the signal of greenhouse warming; however, the evidence is not definitive. There has not been the steady warming this century that might be expected from a steady increase in greenhouse gas concentrations. The record shows rapid warming until the end of World War II, a slight cooling through the mid-1970s, and a second period of rapid warming since then.

Indeed, there is much about which the scientific community is unsure. While there is consensus about the greenhouse effect as a scientific proposition, there is debate as to the precise amount of warming that will occur, the rate at which such a temperature increase might happen, and the regional patterns of climatic change that can be expected from further increases in the atmospheric concentration of carbon dioxide and the other greenhouse gases. The cumulative effects of chlorofluorocarbons, nitrogen oxides, methane, ozone and other trace gases could be comparable to that of carbon dioxide over the next century.

The lack of unanimity of opinion can be accounted for by the difficulties in predicting climate change. In the rest of this chapter I shall outline how climate modelling is done, and its problems. I shall then look at the complications of trying to make the models predict the future. Given these problems, I shall look at how models can be tested and the possible errors in prediction made by those in use at the moment.

CLIMATE MODELLING

Clearly it would help human societies prepare for, or try to stop climate change in the future if we could predict that future in some detail. The processes that make up a planetary climate are too large and too complex to be reproduced physically in laboratory experiments. Fortunately, they can be simulated mathematically with the help of a computer. In other words, instead of actually building a physical analogue of the land-ocean-atmosphere system, one can devise mathematical expressions for the physical principles that govern the system – energy conservation, for example, and Newton's laws of motion – and then allow the computer to calculate how the climate will evolve in accordance with the laws. Mathematical climate models cannot simulate the full complexity of reality. They can, however, reveal the logical consequences of plausible assumptions about the climate. In this context climate modelling is emerging as a field of more than academic interest: it is becoming a fundamental tool for assessing public policy.

Although all climate models consist of mathematical representations of physical processes, the precise composition of a model and its complexity depends on the problem that it is designed to address, in particular on how long a period of the past or future is to be simulated. Some of the processes that influence climate are very slow: the waxing and waning of glaciers and forests, for example, or the motions of the earth's crust or the transfer of heat from the surface of the ocean to its deeper layers. A model designed to forecast next week's weather ignores these variables, treating their present values (the extent of ice coverage, for instance) as external, unchanging 'boundary conditions'. Such a model simulates only atmospheric change. On the other hand, a model designed to simulate the dozen or so ice ages and interglacial periods of the past million years must include all the above processes and more.

Climate models vary also in their spatial resolution, that is, in the number of dimensions they simulate and the amount of spatial detail they include. An example of an extremely simple model is one that calculates only the average temperature of the whole earth. Such a model is 'zero-dimensional'; it collapses the real temperature distribution on the earth to a single point, a global average. In contrast, three-dimensional climate models reproduce the way temperature varies with latitude, longitude, and altitude. The most sophisticated of these are known as general-circulation models (GCMs). They predict the evolution with time not only of temperature but also of humidity, wind speed and direction, soil moisture, and other variables.

General-circulation models are usually more comprehensive than simpler models, but they are also much more expensive to design and run. The optimal level of complexity for a model depends on the problem being addressed and on the resources available to set the model up; more is not necessarily better. Often it makes sense to attack a problem first with a simple model and then employ the results to guide research at higher resolution. Deciding how complicated a model to use for a given task involves a trade between completeness and accuracy versus tractability and economy. This tradeoff is more often a value judgement based on scientific intuition than a judgement based strictly on scientific method.

GRIDS AND PARAMETERS

Even the most complex GCM is sharply limited in the amount of spatial detail it can resolve. No computer is fast enough to calculate climatic variables everywhere on the earth's surface and in the atmosphere in a workable length of time. Instead, calculations are executed at widely-spaced points that form a three-dimensional grid at and above the surface. The model my colleagues and I at the National Center for Atmospheric Research use is typically run with a grid with nine layers stacked to an altitude of about 30 kilometers. The horizontal spacing between grid points is roughly 4.5 degrees of latitude and 7.5 degrees of longitude.

The wide spacing creates a problem: many important climatic phenomena are smaller than an individual grid box. Clouds are a good example. By reflecting a large fraction of the incident sunlight back to space or efficiently absorbing and emitting thermal infrared radiant energy, they help to determine the temperature on the earth. Predicting changes in cloudiness is therefore an essential part of reliable climate simulation. Yet no global climate model now available or likely to be available in the next few decades has a grid fine enough to resolve individual clouds, which tend to be a few kilometres rather than a few hundred kilometres in size.

The solution to the problem of sub-grid-scale phenomena is to represent them collectively rather than individually. The method for doing so is known as 'parameterisation'. It consists, for example, of searching through climatological data for statistical relations between variables that are resolved by the grid and ones that are not. For instance, the average temperature and humidity over a large area (the size of one grid box, say) can be related to the average cloudiness over the same area; to make the equation work one must introduce parameters, or proportionality factors, that are derived empirically from the temperature and humidity data or from a higher resolution process model. Since a GCM can calculate the temperature and humidity in a grid box from physical principles, it can predict through a parameterisation the average cloudiness in the grid box even though it cannot predict individual clouds.

To fully simulate the climate the models must take into account the complex feedback mechanisms that influence it. Snow, for example, has a destabilising, positive-feedback effect on temperature: when a cold snap brings a snowfall, the temperature tends to drop further because snow, being highly reflective, absorbs less solar energy than bare ground. This process has been parameterised fairly well in climate models. Unfortunately other feedback loops are not as well understood. Again, clouds are a case in point. They often form over warm, wet areas of the earth's surface, but depending on the circumstances they may have either a stabilising, negative-feedback effect (cooling the surface by blocking sunlight) or a positive one (warming the surface further by trapping heat).

CLIMATE SENSITIVITY

Uncertainty about the nature of important feedback mechanisms is one reason the ultimate goal of climate modelling – forecasting reliably the future of key variables such as temperature and rainfall patterns – is not yet realisable. Another source of uncertainty that is external to the models themselves is human behaviour. To forecast, for example, what impact carbon dioxide emissions will have on climate one would need to know how much carbon dioxide is going to be emitted.

What the models can do is analyse the sensitivity of the climate to various uncertain or unpredictable variables. In the case of the carbon dioxide problem one could construct a set of plausible economic, technological and population-growth scenarios and employ a model to evaluate the climatic consequences of each scenario. Climatic factors whose correct values are uncertain (such as parameters in the cloud-feedback parameterisation) could be varied over a plausible range of values. The results would indicate which of the uncertain factors is most important in making the climate sensitive to a carbon dioxide buildup. One could then focus research on those factors. The results would also give some idea of the range of climatic futures to which societies may be forced to adapt.

A number of workers have attempted to model the possible climatic impacts of carbon dioxide. Most of them have followed the same approach: they give the model an initial jolt of carbon dioxide (usually doubling the atmospheric concentration), allow it to run until it reaches a new thermal equilibrium and then compare the new climate to the control climate. One of the most widely cited results found that both a doubling and a quadrupling of atmospheric carbon dioxide would produce a summer 'dry zone' in the North American grain belt, but that soil moisture in the monsoon belts would increase. This model reached its new equilibrium after several decades of simulated time.

In reality, however, the approach to equilibrium would probably be much slower. This model omitted both the horizontal transport of heat in the ocean and the vertical transport of heat from the well-mixed surface layer to the ocean depths. Both processes would slow the approach to thermal equilibrium; the real transition would probably take more than a century.

Schneider and Thompson (1981) developed simple one-dimensional models that demonstrated the importance of the transient phase of warming. Regions at different latitudes approach equilibrium at different rates, essentially because they include different amounts of land; land warms up faster than the oceans. Also, oceans have different vertical mixing rates at different locations. Hence during the transient phase, the warming and other climatic effects induced by the enhanced greenhouse effect could well display world-wide patterns significantly different from the ones inferred on the basis of equilibrium GCM simulations. Furthermore, the social impact of climatic changes would probably be greatest before equilibrium has been reached and before human beings have had a chance to adapt to their new environment.

To represent the transient phase adequately one would need to couple a three-dimensional model of the atmosphere with a three-dimensional model of the ocean that included the effects of horizontal and vertical heat transport. A handful of coupled models have been run, but none for long enough to simulate the next century. The coupled models are still too uneconomical for that task, and they are also not yet trustworthy enough. Once they have been improved, one will be able to state with more confidence how the climatic impacts of rising levels of greenhouse gases might be distributed in space and time. Until then one can only cite circumstantial evidence that the impacts are likely to be significant.

VERIFYING CLIMATE MODELS

Perhaps the most perplexing question about climate models is whether they can ever be trusted enough to provide grounds for altering social policies, such as those governing carbon dioxide emissions. How can models so fraught with uncertainties be verified? There are actually several methods. None of them is sufficient on its own, but together they can provide significant (albeit largely circumstantial) evidence of a model's credibility.

One method is to check the model's ability to simulate today's climate. The seasonal cycle is one good test because the temperature changes involved are large – several times larger, on average, than the change from an ice age to an interglacial period. General-circulation models do remarkably well at mapping the seasonal cycle, which strongly suggests that they are on the right track. The seasonal test is encouraging as a validation of 'fast physics' such as cloudiness changes. However, it does not indicate how well a model simulates slow processes, such as changes in deep ocean circulation, that may have important long-term effects.

Another method of verification is to isolate individual physical components of the model, such as its parameterisations, and test them against a high-resolution sub-model or real data from the field. For example, one can check whether the model's parameterised cloudiness matches the level of cloudiness appropriate to a particular grid box. The problem with the former test is that it cannot guarantee that the complex interactions of many individual model components are properly treated. The GCM may be good at predicting average cloudiness but bad at representing cloud feedback. In that case the simulation of the overall climatic response to, say, increased carbon dioxide is likely to be inaccurate.

Overall validation of climatic models thus depends on constant appraisal and reappraisal of performance in such categories. Also important are a model's responses to such century-long forcings as the 25 per cent increase in carbon dioxide and other trace greenhouse gases since the Industrial Revolution. Indeed, most climatic models are sensitive enough to predict that warming of at least 1 degree centigrade should have occurred during the past century. The precise 'forecast' of the past 100 years also depends upon how the model accounts for such factors as changes in the solar constant or volcanic dust. Indeed, as recent data shows, the typical prediction of a degree centigrade warming is broadly consistent but somewhat larger than observed. Possible explanations for the discrepancy include: (i) the state-of-the-art models are too sensitive to increases in trace greenhouse gases by a rough factor of two; (ii) modellers have not properly accounted for such competitive external forcings as volcanic dust or changes in solar energy output; (iii) modellers have not accounted for other external forcings such as regional tropospheric aerosols from agricultural, biological, and industrial activity; (iv) modellers have not properly accounted for internal processes that could lead to stochastic or chaotic behaviour; (v) modellers have not properly accounted for the large heat capacity of the oceans taking up some of the heating of the greenhouse effect and delaying, but not ultimately reducing, warming of the lower atmosphere; (vi) present model forecasts are typically run for equivalent doubling of carbon dioxide whereas the world has only experienced a quarter of this increase, for which it may be a mistake to expect the models to be proportionally accurate; (vii) our inconsistent network of thermometers over the past century may have underestimated actual global warming this century.

Despite this list of excuses why observed global temperature trends in the past century and those anticipated by most GCMs disagree somewhat, the two-fold discrepancy is still encouraging. Most climatologists do not yet proclaim the observed temperature records to have been caused beyond doubt by the greenhouse effect. Thus, a greenhouse-effect signal cannot yet be said to be unambiguously detected in the record. It is possible that the observed trend and the predicted warming could still be chance occurrences. Nevertheless, this empirical test of model predictions against a century of observations certainly is consistent to a rough factor of two. This test is reinforced by the good simulation by most climatic models of the seasonal cycle, and the present distribution of climates on earth. When taken together, these verifications provide a strong circumstantial case that the modelling of sensitivity of the global surface air temperature to greenhouse gases is probably valid within roughly two-fold. Another decade or two of observations of trends in the earth's climate, of course, should produce signal-to-noise ratios sufficiently high or low for almost all scientists to know whether present estimates of climatic sensitivity to increasing trace gases have been predicted well or not. But waiting for such conclusive, direct evidence is not a cost-free proposition: by then the world will already be committed to greater climatic change than it would be if action were taken now to slow the build-up of greenhouse gases. Of course, whether or not to act is a value judgement, not a scientific issue.

THE MODERN GREENHOUSE

Changes in climate on the scale predicted by the models could threaten natural ecosystems, agricultural production and human settlement patterns. For instance, forests probably could not sustain the fast migration required by the projected warming, and many ecosystems cannot migrate at all. Water supplies and quality could be affected by evaporation of moisture, reducing stream runoff. If precipitation declines in arid farm lands, viable acreage could fall by nearly a third. Most workers expect a global temperature increase of a few degrees centigrade over the next fifty or 100 years to raise sea level by between 0.2 and 1.5 meters as a result of the thermal expansion of the oceans, the melting of mountain glaciers and the possible retreat of the Greenland ice sheet's southern margins. (Ice could actually build up in Antarctica owing to warmer winters, which would probably increase snowfall.)

Clearly these direct effects of climatic change would have powerful economic and political consequences. To be sure, not everyone would lose. But how could the losers be compensated and the winners charged? The issue of equity would be particularly thorny if it spanned borders – if the release of greenhouse gases by the economic activities of one country or group of countries did disproportionate harm to other countries whose activities had contributed less to the buildup.

In the face of this array of threats, three kinds of responses could be considered. First, some workers have proposed technical measures to counteract climatic change – deliberately spreading dust in the upper atmosphere to reflect sunlight, for instance. Yet if unplanned climatic changes themselves cannot be predicted with certainty, the effects of such countermeasures would be still more unpredictable. Such 'technical fixes' would run an appalling risk of misfiring.

A second class of action favoured by many economists is that of adaptation. Arguing from the large uncertainties in climate projections and the large investments required to avert outcomes that may never materialise, they suggest merely modifying infrastructure that will need to be replaced in any case, such as water supply systems and coastal structures. This could be done passively as we wait for events to unfold, or actively as we replan systems such as water supply, with a view to changing climate, but so that if climate change does not occur, the system is still useful.

The third and most active category of response is prevention: curtailing the greenhouse-gas build-up, by reducing carbon dioxide emissions and emissions of other greenhouse gases. Such proposals for immediate action are often costly and politically controversial, but perhaps there is some simple principle that can help us to choose which preventive or adaptive measures to spend our resources on. I believe it makes sense to take actions that will yield 'tie-in' benefits, even if climatic changes do not materialise as forecast. For example pursuing and encouraging energy efficiency is economic, saves our resources, and cuts carbon dioxide emissions.

Society is thus faced with a classic example of the need to make decisions with imperfect information. Some projected climatic impacts appear severe; but perhaps these could be mitigated if we know what to expect and if we choose to respond. At the same time, there is the risk of investing resources to prevent an impact that may not appear, or that may appear where least expected. The need to know details about timing and distribution of future climate changes has been stated in many scientific and political fora, and detailed climate impact studies have been commissioned.

As the state-of-the-art of predicting climatic change evolves, scenarios will need to be regularly revised and improved, and the implications for environmental and societal impacts reassessed. Through this iterative process climate scientists hope that clearer understanding of potential climatic impacts will develop. Of course, while scientists study and debate, the world becomes committed to a growing dose of greenhouse gases and their impacts. The rates at which climate change could be evolving are extremely rapid when compared to most paleoclimatic trends. It is questionable whether natural ecosystems and human activities can adapt easily to such plausible rates of change, suggesting some urgency for accelerating the rate at which the scientific community is likely to resolve uncertainties in climatic scenarios and impact assessments.

4 Acid Precipitation

Matthew Wilkinson and Sarah Woodin

Air pollution is not a new problem. London suffered from the smoke pollution of coal burning more than six centuries ago, and in 1273 the first air pollution law was passed to counter the nuisance of smoke from domestic fires in the city. Air pollution became more serious with the intense industrialisation that began more than 200 years ago, and a series of Smoke Abatement Acts were passed in the 1850s designed to control the polluting emissions from alkali works. In 1852 R. A. Smith, one of the first air pollution inspectors, published a scientific paper describing the pollution of air and rain around Manchester, in the north of England. In this he made the comment that 'all the rain was found to contain sulphuric acid in proportion as it approached the town'. Twenty years later he published detailed studies of the polluted state of air and rain and coined the term 'acid rain'.

Smogs, like the one which hung over London for five days in 1952 and during which 4,000 more deaths than usual were recorded, have been associated with industrial emissions for a long time. But it was not until the 1960s that scientists were able to link the dwindling stocks of fish in tens of thousands of lakes and streams in Scandinavia to atmospheric pollution. The threat to their fishing and its heritage came as something of a culture shock to the Scandinavian nations, as did the decline of the forests in West Germany to the Germans, when it became particularly marked in the early 1980s. Likewise, the decline of the sugar-maple and the acidification of the boundary waters has sensitised the Canadians, especially since a substantial proportion of the pollution responsible originates in the United States. In response, to give one small, but pointed, example, the Canadians began to issue fliers to all the car owners from the States in their National Parks, each one particular to the home state of the recipient. The Scandinavian governments have adopted the kind of vigorous and vociferous approach to the scourge that is usually only displayed by environmental pressure groups.

Emissions of sulphur dioxide and nitrogen oxides from industry and vehicle exhausts are the main source of acid pollution. When these compounds dissolve in water droplets in the atmosphere they form the powerful sulphuric and nitric acids. With their precipitation in rain or snow, these acid pollutants make contact with surfaces – this is known as 'wet' deposition. Wet deposition also describes the impact of pollution from cloud or fog, although this is sometimes called 'occult' deposition. Wet deposition can be carried over great distances by the wind before descending. Alternatively, the gaseous emissions may be deposited directly onto surfaces, in which case it is described as 'dry' deposition.

A property of all acids is their ability to release positive hydrogen ions (H^+). Acidification amounts to an increase in the concentration of hydrogen ions in soil and water. Chemical and biological processes, both in soil and water, are greatly affected by a change in the levels of hydrogen ions. There are natural mechanisms to oppose and neutralise acidification: for example, in a lake these so-called buffer actions are performed by carbonates (e.g. HCO_3). Lakes in lime-rich areas receive a plentiful and continuous supply of calcium carbonate – lime – and therefore run far less risk of becoming acidified. Several buffering processes operate in soil, the most important being the weathering of various minerals in the rock. So long as lime-based minerals abound in the soil, its pH value will stay high and the risk of acidification low. Should the pH value drop through an acid deposition, a major buffering process is set into action which releases aluminium compounds in the soil. Free aluminium in soils is known to damage tree roots. Aluminium can also infiltrate groundwater and surface water, and so intensify the harm to fish populations of streams and lakes already afflicted by acidification.

The typical signs of an acidified lake are the disappearance of fish and the reduction in the numbers of local plant and animal species. The water loses the opaque quality endowed to it by the suspension of plankton and turns 'beautifully' clear. Rosette species such as lobelia, quillwort and shoreweeds along the water's edge become overwhelmed by bog moss, which thrives in acid water. Pondweeds also die out, although waterlilies survive, since their roots take nourishment from the sediment layers of the lake-bed beyond the reach of acidification. After a time the fish will have vanished completely, bogmoss will have spread out to cover the majority of the lake floor and few plants or animals will remain. In the absence of the fish and their predatory ways, some insects, such as the water beetle and the dragonfly, multiply in numbers.

The acidification of lakes and streams is inseparable from that of the soil, for the simple reason that 90 per cent of the water that they hold has passed through the ground. Several acidifying processes take place naturally in the soil, most prominently, the uptake of nutrients by plants. Plants compensate for the intake of positively charged nutrient ions by releasing positively charged hydrogen ions. Therefore growth itself is acidifying, but in an ecosystem where growth and decay are in equilibrium there will be no net acidification. Until ten years ago scientists in general held the belief that the buffering ability of the soil could to a great extent counteract acidification from acid deposition. This is now known to be untrue. Throughout much of southern Scandinavia the pH value of soils has fallen by between 0.3 to 1.5 units and the soil in badly stricken parts has been soured to a depth of a metre. In parts of West Germany and Austria the pH value has dropped by similar amounts from its level of two or three decades ago.

Once the soil has been contaminated, the populations of earthworms and bacteria decline and are succeeded in their function as the decomposers of dead matter by fungi. Fungi break down matter more slowly and so release fewer nutrients to soil that already suffers from a poverty in nutrients – one reason for its original susceptibility to acidification. Soil types vary in their susceptibility for and sensitivity to acidification. For example, the podzol forest soils, that sustain both conifers and deciduous trees, are gravely threatened by it. These soils are most common in areas of heavy rainfall and, because they are coarse-grained and poor in lime, are victim to the debilitating leaching of nutrients by acid rain. Brown soils, usually found in areas of meadowland and broad-leaved forest, contain more nutrients than podzols but suffer particularly from aluminium poisoning in the event of acidification.

Deposition from the atmosphere is not the primary cause of the acidification of farmland, of which modern agricultural techniques account for between 15 and 50 per cent. A vast number of hydrogen ions are released into the soil when the ammonium in artificial fertiliser is converted into nitrates. Air pollution can, however, affect crops directly and markedly. Ozone, in particular, damages many crops and reduces yields. According to estimates made in the United States in the early 1980s, each year ozone inflicts a harvest loss worth between two and four billion dollars. Studies carried out in Sweden show that yields of potatoes, peas and spinach fell by between 10 and 30 per cent, in the conditions of ozone concentration common in that part of the world. (It should be noted that ozone in the upper atmosphere – the ozone layer – is naturally formed and protects the earth. Ozone in the lower atmosphere is a pollutant, made by a reaction between nitrogen dioxide, hydrocarbons and light.)

That forests are damaged by air pollution is beyond dispute. In 1985, over half the forest area of West Germany, which totalled 3.8 million hectares, was found to have been damaged. In Czechoslovakia 200,000–300,000 hectares of high-lying forest has died. A damaged tree can exhibit a number of symptoms, and it is often extremely difficult to establish a link between any one type of injury and any one cause, whether it be air pollution, climatic strain or insect attack. One stress may sometimes prompt a range of disorders. Air pollution can predispose trees to damage by other factors, inflict damage directly or aggravate the symptoms of an already injured tree.

'Dry' or 'wet' deposition can directly harm a tree's needles or leaves, when it penetrates its protective layer of wax. Air pollution can cause the malfunction of the stomata – the minute openings which regulate the evaporation of the water within a tree. It can weaken or break the membranes inside leaves or needles, which then leak out nutrients and water. Trees incur damage indirectly with the consequent decrease in the supply of nutrients in the ground after acidification and the release of poisonous substances, such as aluminium, as mentioned above. A tree's roots may become incapable of absorbing the required nourishment and water. Disruption of symbiosis of root hairs with fungi impedes the vital exchange of nutrients between tree and fungi. In this enfeebled conditon a tree cannot survive disease, attack from insects and other parasites, hot summers or severe winters.

Trees vary in sensitivity to polluted air. Damage to conifers may take the visible aspect of the loss, yellowing or browning and shortening of needles, the growth of adventitious shoots, loss of root hairs and stunted root growth, narrower annual rings, more ruptures in the trunk, drooping branches and the thinning of the crown. Deciduous trees may also develop cracks and tumours in the trunk, and suffer stunted shoot formation. Of course, a plethora of flora and fauna is also harmed by acid deposition. Frogs, which lay their spawn in the shallow waters alongside lakes and ponds, have their breeding habits disrupted. Small birds living by acidified waters have reproductive problems. For example, the eggs of pied flycatchers and willow warblers have sometimes been found to be too thin for successful hatching due to the consumption of aluminium by the insects that they eat.

It is not likely that acid groundwater will itself harm humans. But where the water is highly acidic, metals such as aluminium and cadmium form in potentially dangerous quantities. Cadmium is the most mobile of ordinary heavy metals and, since it is one of the elements to which people in industrialised countries have already been exposed in dangerously high concentrations, we need to be alert to the danger of cadmium levels rising in groundwater. Cadmium collects in the renal cortex where it causes lesions. Smokers, who already ingest large amounts of cadmium, may be in great danger if amounts of it increase in drinking water. Increased doses of aluminium in water are suspected of causing both Alzheimer's and Parkinson's diseases, both of which are terminal and turn the sufferer prematurely senile.

Sulphur dioxide and nitrogen oxides are corrosive agents and many act synergistically in acid rain to corrode man-made and natural materials. Steel, whether uncoated, painted or galvanised, zinc, sandstone, limestone, plastics, paper, leather, and textiles are all prone to corrosion. Pipes, steel structures and even concrete foundations can be affected by the acidification of soil and groundwater. Homes and institutions may become insanitary when acidic water pits their pipes and leaves deposits in which bacteria flourish. The disfigurement to ancient monuments and other cultural treasures that acid rain has brought about has concerned many people for some time.

The direct effect of polluted air to the health of humans ranges from the trifling to the very disturbing for those with breathing disorders e.g. asthmatics, the elderly, and sufferers from heart and circulatory diseases. The immediate effect of breathing air that has suddenly become highly polluted may be irritation of the respiratory tract or even the contraction of a breathing disorder. In the Ruhr, West Berlin and in Mexico City smog has sometimes been so toxic that citizens have been advised to stay indoors. Breathing polluted air consistently over a long period exposes the individual to the possibility of chronic illness, with recurrent respiratory trouble and even damaged organs.

Air pollution and acid precipitation can be mitigated in the short term. The acidification of lakes and waterways can be neutralised by liming. This process involves the application of crushed limestone to the surface of the water, which raises the pH of the water and causes the aluminium and other metals to sink to the bed of the lake. After liming many species of flora and fauna return to the lake and fish can once again reproduce. However, liming is far from the simple panacea that it is sometimes made out to be and the little-studied side effects of liming worry many conservationists. Lime is a very important ecological determinant, and the plant and invertebrate communities of lime-rich and lime-poor areas are very different. Freshwater acidification is most severe in areas where the soils are poor in lime. It is these areas to which lime is most rigorously applied, and liming does not, therefore restore pre-acidification conditions, but creates new ones. The ecosystems of a lake or river that have already been degraded by acidification may be upset further by liming. The liming of catchment areas is even more worrying and is becoming more common. In these areas it appears that the populations of invertebrates have been reduced and small mammals, such as shrews, destroyed. Many catchment areas are, indeed, noted for the diversity and rarity of the species that they harbour. Even in lakes, species whose habitat is on the lake floor may be killed off by the precipitation of aluminium after liming. Along with these ecological drawbacks come practical ones. In rivers, streams and lakes with a rapid turnover of water, liming must be carried out frequently. This is very expensive. It is the enormous cost of liming that prevents its widespread use.

The only long-term solution to the problem of acidification is the reduction of emissions of pollutants. About 70 per cent of the emissions of sulphur dioxide and 40 per cent of nitrogen oxides comes from the combustion of oil and coal in power plants. There are well-established methods for limiting these emissions before, during and after combustion. Fuels with a low sulphur content can be burned – there is some coal on the market which contains less than 0.5 per cent sulphur. But the supply of fossil fuels with a low sulphur content is limited, which promotes the need for technical desulphurisation.

One method of desulphurising coal is to grind it into powder and then separate the light and the heavy fractions. This permits the removal of up to half the sulphur. Oil can also be desulphurised in this way. Sulphur dioxide may be retained from combustion by the addition of limestone or slaked lime, when sulphur oxides react with the lime and are removed after the reaction in the form of ash. This method works in a conventional combustion plant or in a fluidised bed where the fuel is mixed with sand and suspended by a stream of air to increase the reaction surface. Combustion takes place at a much lower temperature in fluidised beds than in conventional plants which means that emissions of nitrogen oxides as well as of sulphur dioxide are diminished. Considerable advances have now been made in the development of methods for the desulphurisation of the fluegases as an alternative procedure. This is more reliable and costs less. One example is the so-called wet/dry technique by which the acid smoke is sprayed with lime sludge and the sulphur is then separated as a dry powder. This process, as well as many others that can be used in existing plants, reduces emissions of sulphur by nearly 90 per cent. According to an OECD study made in 1981, the discharges of sulphur in Europe could have been halved in ten years if techniques then available had been implemented. The corresponding increase in energy costs would have been a mere 3 per cent. One of the purposes of this OECD study was to confound those who refused to apply less polluting methods because of overriding economic pressures. Of course, if the relevant authorities and consumers were prepared to pay for it, desulphurisation could be speeded up. Nitrogen oxide emissions from power plants can be brought down by anything from 20 to 40 per cent with more efficient combustion. The Japanese have a catalytic method of combustion which reacts the nitrogen oxides with ammonia and so converts them to nitrogen gas and water. This method is 80 per cent efficient.

From between 40 and 60 per cent of nitrogen oxide emissions in Britain, for example, are generated by motor vehicles and, despite the disparity of their numbers, petrol and diesel-fuelled vehicles share the responsibility for this figure equally. There is a great variety of possible ways to act on this front. These include the use of more efficient engines, the slower driving of older cars, the imposition of speed limits, the reduction of road traffic by fuel or driving taxes and the mandate for vehicles to be fitted with catalytic converters. Some of these measures are perceived in varying degrees to be challenges to personal freedoms, which may have to be reappraised in our attempts to tackle this ecological threat. The emissions of pollutants can, however, be most effectively limited by the conservation of energy. The opportunities for cutting energy consumption are almost endless both industrially, institutionally and domestically. In theory the energy consumption of Sweden could be halved in the next twenty-five years *and* production increased by 50 per cent, with the combined application of available technologies, and energy conservation measures.

A number of calculations that were intended to determine acceptable levels of acid deposition have all produced approximately the same result. An annual fallout of 0.5 of a gram of sulphur per square metre is the most that sensitive areas can withstand. Just over two grams of sulphur per square metre now fall over southern Scandinavia every year. Assuming a consistent, natural level of 0.1 to 0.2 of a gram, emissions need to drop by at least 80 or 90 per cent.

It is no accident that the problems of acidification were first noticed in Scandinavia. It has, by and large, a thin soil cover and a bedrock that is deficient in lime, and so is sensitive to acid rain. Moreover, the prevailing westerly and southwesterly winds of northern Europe carry air pollutants over Scandinavia from Britain and the continent. However, very few countries in the northern hemisphere, if any, can remain untouched by the problem, whether as a net 'importer' or 'exporter' of air pollution.

The history of international cooperation in the confrontation of acidification is somewhat chequered. At the United Nations Conference on the Human Environment in Stockholm in 1972, the Swedish government presented a report on transnational, airborne pollution. In this report the diminishing growth of Swedish forests was predicted but treated sceptically. The conference did, nevertheless, declare the principle that nation states were duty-bound to ensure that the activities of their country did not threaten the environment of another's. The first international conference on the effects of acidification took place in Norway in 1976, and the following year Norway proposed the setting up of a convention to restrict discharges of sulphur dioxide.

At a meeting of the EEC Council of Ministers in Geneva in 1979, thirty-four countries from Europe and North America signed a treaty, known as the EEC convention for the control of the transboundary transport of polluted air. But it was not until December 1982 that it had been ratified by enough states to bring it into force. Once again, this did not make the control of emissions mandatory. It stated no more than the principle that the signatories would 'endeavour to limit air pollution, including long-range transboundary air pollution'. In 1983 the Scandinavian countries proposed the limitation of emissions to a level cut by 30 per cent. The proposal was supported by Canada, West Germany, Switzerland and Austria, which, together with the Scandinavians, were christened the 30 Per Cent Club.

In the summer of 1984, ministers representing all the signatories to the EEC Convention met in Munich and decided that a legally binding international agreement to reduce sulphur emissions by 30 per cent was needed. A year later in Helsinki the environment ministers of twenty-one countries signed such a document. Fouteen other countries party to the Convention did not sign, among them Poland, Britain and the United States. Economic pressures were responsible for Poland and several Southern European countries omitting to sign. But the United States and Britain, two vast net exporters of air pollution, insisted that more research was needed into the spread of pollution before they would agree to take action. Moreover, Britain maintained that both the proposed percentage of reduction and its timetable were arbitrary and unscientifically calculated. This hesitation by the British government was regarded by some of the signatories as stalling goaded by national obstinacy towards acting under international pressure. They found it particularly irksome both ecologically and emotionally, since Britain is the single largest discharger of pollutants in Europe, barring the Soviet Union.

In 1985 the EEC member states agreed to tighten emission standards for petrol-driven cars. Denmark, however, dissented, appealing for more stringent regulations and Britain has refused to abide by any rules that necessitate the use of catalytic converters. Nevertheless, despite this political and diplomatic sparring, emissions of sulphur dioxide in Europe have decreased in the last decade by about a quarter, as a result of EEC Directives on the reduction of emissions from large combustion plants and vehicles. In Canada, a country that suffers badly from the effects of acidification, strict measures are being imposed to reduce industrial emissions. In the United States government action has been slow.

Potential acid pollution from further industrialisation in the developing world merits at least as much consideration as the prevention of further environmental damage in the developed world, and recommends the sharing of technologies by the 'North' with the 'South' very forcibly. Recent events also present the opportunity for Western Europe to assist in improving the conditions of appalling air pollution in much of Eastern Europe.

5 Deforestation: a Botanist's View

Ghillean T. Prance

Deforestation has become a familiar term in the latter half of the 1980s, and many people react as if it were a new discovery. In fact it has been a problem since the origin of agriculture when man began to clear the forests to plant crops. This in turn enabled him to enlarge the population size and therefore, the tendency to deforest has continued and increased. Rather than turn straight to the forests of Brazil or Malaysia, we need to begin nearer to home and realize that when the Romans reached Britain it was largely a forested island, and that the scenery which we think of as rural today with its farmland, hedges and downs, is artificial. Hannibal did not have to travel to Central or Southern Africa to obtain elephants for his famous attempt to cross the Alps because these animals roamed the forested areas of North Africa just across the Mediterranean Sea from Italy. Israel is a semi-desert country because it was deforested in Biblical times. The result of forest loss in the Middle East and North Africa is obvious, the increase of desert.

Deforestation is a complex issue because of the difference between the developed and the developing world. In recent times clearing of forest has slowed down and even been reversed in much of the developed world, yet it has accelerated in the developing world, especially during the last two decades. Deforestation has therefore become a North versus South issue. Even in the North the statistics can be confusing. Although 239,000 km^2 of farmland in the United States was converted back to forest between 1910 and 1979, there was a net decrease in forest of 254,000 km^2 between 1974 and 1984. This was due to forest being cut both for new farmland and for urban, industrial, mining and other uses. 87 per cent of the rainforests of Washington State and Oregon have been logged. In India on the other hand there was a 14,000 km^2 increase in forest cover for the same period.[1] China has suffered extremely badly from deforestation and large areas of the southern region have become eroded and desertified. Yet between 1974 and 1984 forest cover in that country increased by 19,700 km^2. In fact, China has replanted about 15,000 km^2 annually since 1965.

Deforestation is included as a topic for this volume because, in spite of the efforts of some countries to reforest, it is still a serious threat to the environment. It is a cause for concern because the rate of deforestation in the tropics has accelerated so rapidly during the last decade. A 1982 estimate by FAO showed that about 7.5 million hectares of tropical closed forest are cleared each year and another 3.8 million hectares of open forest are also cleared, making a total of 11.3 million hectares a year.[2] This was thought by some people to be an exaggeration of the rate of deforestation, but more recent statistics, based on remote sensing, have proved the FAO figures to be conservative. For example, Malingreau and Tucker showed that between 1982 and 1985 over 9 million hectares were deforested in the Brazilian states of Rondônia and northern Mato Grosso.[3] 1987 was a peak year for deforestation in Rondônia and other areas in the southern part of the Brazilian Amazonia. Malingreau and Tucker showed that 60,000 km^2 of Rondônia was deforested that year, another 65,500 km^2 in Mato Grosso and 5,300 km^2 in Acre giving a total of over 12 million hectares in three states of Brazil alone.[4] The deforestation in the western Amazon for 1987 exceeded the worldwide total of the FAO estimate for 1982! These statistics are enough to give rise to grave concern without even considering what is happening in Indonesia, Madagascar and other parts of the tropics.

Before the dawn of agriculture, closed forest covered some 46.28 million square kilometres of our planet and this has now been reduced to 39.27 million km^2 while open woodland has declined from 15.23 million km^2 to 13.10 million km^2.[5] This destruction has already caused grave environmental damage affecting climate patterns, the gaseous contents of the atmosphere and species loss. Deforested areas have turned to desert and to dust bowls, and it is vital to respond to this crisis before it is too late.

Deforestation today is taking place largely in the tropics which is cause for much greater concern for two main reasons:

1. Most tropical rainforests grow on extremely poor soils that become unproductive once the forest is removed.
2. The tropical rainforests, although they cover only 14 per cent of

the land surface of the globe,[6] harbour over 50 per cent of the biological species.

RICH FORESTS – POOR SOIL

Most of the area covered by tropical rainforest is on extremely poor soil, latosols or ferralsols and the even more impoverished podzols. The forest is not luxurious and of high biomass *because of* the soil, it grows there *in spite of* the soil. The nutrients of the ecosystem are in the trunks, branches and leaves of the trees, and in the fauna rather than in the soil. When the forest is cut and burned these nutrients are released in the form of ash. This allows good growth of replacement crops for the first year or two. However, another property of the clay latosols is that they do not have the colloidal properties to hold the nutrients. The heavy tropical rains wash these precious minerals and salts away into the streams and rivers leaving an even poorer ecosystem. This accounts for the large amount of abandoned cattle pastures along the Transamazon Highway of Brazil and the extensive areas of the useless ylang-ylang (*Imperata*) grasslands in formerly forested areas of the Malaysian tropics. If sustainable management of the tropics is to be achieved, careful control of the form of use in relation to the soil type must be exercised. In the rain forest there are smaller areas of richer soils such as nitosol or the gleysols of flood plain areas that are suitable for agriculture. The tragedy is that much of the tropical deforestation has occurred on areas of poor soil for short-term profit, or even losses that can be deducted from tax, rather than because of soil potential.

TROPICAL FOREST DIVERSITY

I was not surprised when a forest inventory which I carried out in 1975 of a hectare of forest near to Manaus, Brazil showed that there were 179 species of trees of 15 cm diameter or more in that small area.[7] The richest known forest in the world is at Yanamono in Amazonia Peru where Alwyn Gentry found 283 species of trees and 17 of lianas of over 10 cm in diameter in a single hectare of forest. For the botanist trained in the temperate region, where there are often only three or four species of trees on a hectare, it is hard to imagine the extraordinary diversity of tropical rainforest. Instead of a familiar uniformity of oak or pine trunks, every tree trunk in a tropical forest looks different from its neighbour. This diversity is not confined to trees. Gentry and Dodson found 365 species of flowering plants in a tenth of a hectare in the Río Palenque Reserve in Ecuador.[8] That is over 20 per cent of the British flora in as little as 0.1 ha. More species of birds, primates, mammals and freshwater fish occur in tropical rainforest areas than anywhere else. Terry Erwin estimated that there are 30 million species of insects in tropical forests.[9] To date scientists have described and named only 1.4 million species of living organisms. The forest is being destroyed before we have catalogued more than 5 per cent of the species.

THE CONSEQUENCES OF DEFORESTATION

Species loss

By far the greatest concern about deforestation, especially of the tropics, is the consequent loss of species. Because of its diversity and the sparse populations of each species within the forest, and because of the amount of local endemism, when the forest is cut, the species of plants, birds, insects and other animals are especially prone to extinction. In certain areas species extinction is already occurring.

The Pacific coastal rainforest of Ecuador, rather isolated from other rainforests, contained between 40 and 60 per cent endemic species of plants.[10] Since 1960 these forests have been felled to be replaced by banana and oil palm plantations. Gentry observed the fate of one of the last forested ridges in western Ecuador, the Centinella Ridge which contained about 100 endemic species of plants that were known to survive only there. Between 1980 and 1984 the area was completely deforested for agriculture and so 100 ridgetop species of plants and certainly many insects became extinct. The new species of *Licania* that I am currently describing from material collected by Gentry is probably already extinct.

Mori and colleagues found that 53.5 per cent of a sample of 1245 tree species from the Atlantic coastal forests of Brazil were confined to these forests.[11] Also, many species of this narrow belt of rainforest parallel to the Brazilian coast are of extremely restricted range. Brown (1979) indicated five centres of endemism for the butterflies he studied within the Brazilian coastal forests. All biologists who have studied the species of plants and animals of these forests have commented on their high level of endemism.[12] Mori and Dean have catalogued the destruction of the Atlantic Coastal forests of Brazil, and today less than 5 per cent of the original one million square kilometres remains intact.[13] This deforestation, performed to make space for cocoa plantations, sugar cane fields and agriculture, has certainly caused extensive extinction of the species that are endemic to these forests.

Many other examples of areas of rainforest with high species endemism and suffering excessive deforestation could be described in detail. Similar extinctions have occurred in such places such as Madagascar, Java, Hawaii and many other tropical islands. The conclusion of various scientists is that we may already be losing as many as six species an hour, and that by the year 2000 between 25 and 50 per cent of all living species may become extinct.[14]

The loss of any single species can have a serious effect on others because all ecosystems are woven together by a complex mass of interactions in the interrelationships of pollination, dispersal, defense from predators, mycorrhiza to help the absorption of nutrients, and many other symbioses. The removal of one link in the chain can have a serious effect on all the others and so a chain reaction of extinctions is set in motion. For example, the Brazil nut flowers will not produce the nuts that are the seeds for the next generation without large euglossine or carpenter bees to visit their flowers.

The consequence of species loss is hard to estimate, but it severely limits our options for the future management and conservation of our planet. Tropical forest plants are a rich source of useful products. Some such as rubber, coffee, quinine and chocolate are well known, but others are as yet unknown: medicines, foods, flavours, fragrances, oils and resins that will be of great potential use if the species in which they occur are not allowed to become extinct.

A study of four tribes of Amazonian Indians showed the extent to which these forest people depend on the rainforest trees for their survival.[15] The Ka'apor Indians of Brazil used 76.8 per cent of the 99 species that occurred on the hectare of rainforest studied and the Chácobo Indians of Bolivia used 74 (or 78.7 per cent) of the 94 species on the study plot.[16] Uses for many rainforest plants will eventually be discovered and many of these are suitable for use in sustainable systems of forest managment. The challenge facing us is to create systems of extraction of products from the forest and of sustainable agroforestry before too many species are lost forever. The more species we lose, the harder it will be to sustain permanently those which survive the initial spate of extinctions.

Many of the useful species that we already know have their wild relatives in the forest and these often contain essential genetic material for the future survival of the crop. For example, there are over twenty other species of the cocoa genus *Theobroma* in the rainforests of tropical America. These contain properties such as resistance to disease that could ensure the future of the cocoa industry. The recently discovered wild species of maize, *Zea diploperennis*, from Mexico is not only perennial but is also resistant to several of the common diseases of corn. Plant breeders will certainly use these properties of the wild species to breed new disease resistant strains. A wild species of peanut from the Amazon was resistant to peanut leafspot and that single disease-resistant gene is estimated to be worth $500 million a year from the benefits it has given to the peanut industry.[17]

Soil loss

Studies from all forested regions of the world show that deforestation causes soil loss. The forest protects the soil from erosion because it acts as a filter to allow the rainfall to reach the soil gradually through the foliage and by slow flow down the tree trunks. In undisturbed tropical rainforest the streams that leave the forest contain almost pure rainwater that is comparable to distilled water in its mineral content. As soon as the forest is cut the mineral content of the streams rapidly rises as the previously closed system of nutrient cycling is disrupted and nutrients escape. This not only causes loss of the valuable and scarce nutrients in the soil, it also results in massing of silt. Many of the species-diverse coral reefs in the tropics are being destroyed by silt caused by deforestation of the watersheds that drain into the sea. The Panama Canal system is threatened by the increased siltation caused by deforestation of the watersheds that drain into this system of lakes. An alarming account of the gravity of soil erosion was given by Brown and Wolf.[18] Soil loss is particularly dramatic when deforestation has taken place on steeply sloped areas. This has been a particularly severe problem in Java and in the Andean region of South America, as well as in Madagascar.

Climate change and desertification

There is a clear link between deforestation and environmental change, although much misinformation has been perpetrated here. For example, tropical rainforests are not the source of much of the world's oxygen. They in fact consume as much oxygen as they produce because they are in equilibrium. However, deforestation followed by burning does release carbon dioxide into the atmosphere and thereby contributes (perhaps 20 per cent) to the increase of atmospheric CO_2, one of the greenhouse gases that is causing the global temperature to rise.

Much more important is the relationship between rain forest and rainfall. Villa Nova and colleagues estimated that 54 per cent of the total rainfall in Amazonia is derived from the evapotranspiration by the trees in the forest.[19] To remove the forest, and thus reduce leaf area, would significantly decrease evapotranspiration and therefore rainfall. A reduction in rainfall leads to changes in the vegetation and so deforestation gradually leads to the increase of more arid vegetative types such as cerrado (savanna scrubland) and caatinga (semi desert).

In areas of tropical dry forest, such as in Central Africa and Northeast Brazil, the deforestation is taking place in drought-prone areas and desertification can easily follow. The problem of drought in the Sahel region of Africa would have been far less bad during the natural dry cycles of climate if the dry forest had been left standing to protect the soil and to increase the rainfall. The deforestation of the species-rich tropical rainforest has drawn much attention because of species loss and culture loss of indigenous peoples. However, we should be equally concerned about less emphasised dry forests because of their vital ecological role in the prevention of desertification. Dry forests have been subjected to deforestation and overuse for a much greater time period than rainforests. They are losing many species due to the overuse of land and the overgrazing by domestic animals, especially goats. It is in the arid lands that were formerly covered by deciduous and semi-deciduous forest types that desertification is taking place. In our quest to save the tropical rainforest we should not lose sight of the importance of tropical dry forests. Daniel Janzen has been campaigning for the tropical dry forests in many places.[20] He has managed to preserve and restore one in Guanacaste, Costa Rica. However, we should also be concerned with the arid forests of northern Venezuela, northeastern Brazil, Central Africa, western Madagascar and Australia, to name a few regions where the threat of desertification is imminent.

Cultural loss

We have already seen above the extent to which the rainforest peoples use the forest trees. Their management techniques are of equal interest for agriculture, but deforestation is unavoidably accompanied by cultural destruction. Many forest peoples have already become extinct and others have been so badly acculturated that they no longer know how to survive in the forest.[21] The understanding of forest cultures is crucial to our ability to create sustainable management systems for the forest. Yet already only a small remnant of these peoples is left to instruct us. Their preservation is vital. I will conclude with a story from my own rainforest experience that shows the lessons from indigenous culture.

I remember well my first day of botanical field work in the rainforests of Surinam. After a long journey by air, two days by canoe and then travel on foot, we reached the area where our expedition was already at work. Near the camp was a huge flowering tree of *Licania* in the plant family Chrysobalanaceae. I had recently completed my doctoral thesis on this subject and was specially eager to see one and to collect dried herbarium specimens from it. Since the tree was far too large to climb, the expedition leader immediately asked one of our most willing bush negro helpers to cut down this forest giant in order that we might collect a few flowering twigs for specimens. The bush negroes are escaped African slaves who rebuilt tribal life in the forest of Surinam and have maintained many of their own traditions. We were all surprised when Frederick refused to cut down the tree. After much coaxing he agreed to cut it down, but only after he had appeased the 'bushy mama', his god. After chanting a prayer and an offering to the bushy mama, Frederick began to cut down the tree, continuing to sing all the time in a loud chant that would show clearly that he was not cutting the tree willingly but that it was the white man who had ordered him to do so and who should be blamed for this unnecessary destruction. When the people cut a tree to build a house or to obtain wood for their famous carvings they will also obtain permission and appease the bushy mama. The trees of the forests have a spiritual value to the people and are not their own property to destroy. This attitude leads to better protection and a more prudent use of resources than in our society. It is an experience which I have never forgotten, and is one of the reasons why I either climb trees or use local tree climbers to collect my plant specimens!

6 Agricultural Pollution: From Costs and Causes to Sustainable Practices

Jules N. Pretty

THE NATURE OF AGRICULTURAL POLLUTION

It has long been known that pollution threatens our environment in many ways, causing damage to natural resources and harming humans. But the link between cause and damage is usually complex and difficult to assess. This is particularly true of agricultural pollution, which is normally diffuse by nature, arising at low levels but leading to significant cumulative impacts. Agricultural pollution is significant for a number of reasons:

- It is a relatively new phenomenon, arising primarily out of the intensification of agriculture to feed growing numbers of people.
- The environment functions as an integrated and interdependent whole, and agricultural pollution inevitably has a wide range of impacts upon the component parts.
- Accurate assessment of the problems and costs of pollution and the development of policies and technologies to reduce pollution, require a partnership between specialists from a wide range of disciplines and institutions, and the people who produce, or suffer the impact of, the pollution.

THE SOCIAL COSTS OF AGRICULTURAL POLLUTION

Agricultural activity can lead to widespread contamination of water, food and air (Table 1 p. 69). Many of these consequences have now been identified in industrialised countries but they are by no means restricted to only these economies. The developing countries are increasingly likely to suffer too.

Although some agricultural pollutants damage agriculture itself, as when pesticides kill natural enemies of pests or contaminate groundwater used for irrigation, mostly they affect other sectors of society. Each impact incurs a cost, which must be paid for by someone. Farmers who do not perceive pollution as affecting their livelihoods can ignore these costs, and as a result pollute more than if they had to bear all the social costs of their activities. Until now, these costs, if recognised, have been taken to be small. This means that agricultural performance, measured in terms of food or fibre production, is accorded a level of success that does not reflect the true internal and external costs.

In practice the assessment of these effects and external costs is very complex, and we must be cautious for two reasons. First, agriculture is not solely to blame. Take nitrates for example: in drinking water they threaten the health of infants by indirectly inducing the blue-baby syndrome. However, they are just as likely to have been derived from human organic wastes as from fertilisers.[1] Pesticides derived from disease vector control programmes as well as agriculture are also potential contaminants of water and food in developing countries. Another example is soil erosion which in some circumstances can be as high from naturally vegetated as from cultivated land.

Second, the level of uncertainty over the route from pollutant to consequence is still high in several cases. For example, the link between nitrates and cancer is still contentious: although the carcinogenic N-nitroso compounds can be synthesised in humans, it is still not known whether this results in cancer. At macro level the epidemiological evidence is also conflicting, and despite many studies no conclusions have yet been reached.[2] Another case concerns pesticide residues in food: that pesticides occur in foods, sometimes in excess of recommended concentrations is known, yet to date there appears to be no evidence of harm arising from their consumption.

Despite these cautions, there is still concrete evidence of disruption, degradation and contamination of water, food and air by agricultural pollution in industrialised and developing countries. Put together they represent very damaging cumulative stresses that could eventually threaten the whole base for agricultural production itself.

UNDERLYING CAUSES

Although each pollution problem has its own peculiar underlying causes, there are a number of common themes that can be drawn out of the complexity. These relate to the trend of agricultural intensification, the relative costs of inputs and economic incentives for overuse, and the perceptions of the people involved in production and consumption: the farmers may produce pollution and by their activities reduce it, and the consumers may affect the market demand for products by their choices.

Intensification of agriculture

For centuries farming was in stable harmony with the environment. Crop residues were incorporated in the soil or fed to livestock, and manures returned to the land in amounts that could be absorbed and utilised. Trees and shrubs were used for wood and livestock fodder, harboured natural enemies of pests, and the roots bound the soil against erosive forces. This kind of traditional mixed farming system was closed and sustainable, generating few external impacts and using few external resources.

But much has changed over recent years. As farms in the industrialised countries and resource-rich areas of the developing countries have become more productive, so they have become larger and fewer in number, increasingly mechanised, less reliant on human labour, more dependent on synthetic fertilisers and pesticides, and less diverse in their range of activities. In the developing countries, these changes have been called a 'Green Revolution', in which the single goal of increased productivity was achieved largely through the introduction of widely-applicable modern varieties of crops augmented by the use of fertilisers and pesticides.

Increasing specialisation has meant that crop residues and livestock manures, once recycled, are now wastes requiring disposal. Intensification has also brought new practices that have in turn led to greater chemical use. In the irrigated lowlands of Asia, the new varieties of rice that are directly sown require more herbicides because weed problems increase when rice is not transplanted. Increased use of nitrogen has increased the incidence of the disease sheath blight, and staggered planting has increased the incidence of other pests. In the UK increased mechanisation in a vegetable region close to London gave an economic advantage to large farmers, who simplified their production system by growing only three families of vegetables. This led to decreased crop rotations and increased use of pesticides on the new higher-yielding varieties.[3]

Even with careful use of fertilisers and pesticides, the potential for pollution of the environment grows with increasing use. In temperate regions some 30–50 per cent of applied nitrogen in fertilisers is lost to the environment and losses are even greater in the tropics.[4] Some of the pesticides too are lost to the environment, or remain in the crop as a residue. For pesticides this is particularly problematic because few are selective. Typically, they act by interfering with basic biological processes that are common to a wide range of organisms. It is theoretically possible to design selective products, but costly. A pesticide manufacturer must now investigate some 23,000 chemicals for each pesticide discovered – thirty years ago the average was only 1800. Producing a compound also costs some $30 million and takes eight to ten years.[5] Thus most new pesticides will be broad spectrum compounds acting on a range of different pest organisms, and inevitably affecting a wide range of non-target organisms, including humans.

In future, much will depend upon the methods of crop fertilisation and pest control that are adopted, and the picture may well change as fertiliser and pesticide usage increases. There has been a dramatic increase in fertiliser consumption over the last fifty years, and now average rates to most intensively farmed arable land are of the order of 120–550 KgN/ha. In the developing countries average rates are still a great deal lower, even though total consumption has grown by 250 per cent over the past ten years.[6]

For pesticides the picture is different: consumption is currently declining in many industrialised nations, and future growth is expected to occur largely in the developing countries. Like fertilisers this is mostly concentrated on the best quality lands. In Tunisia, for example, almost all the pesticides are used on the irrigated lands that comprise only a small fraction of total agricultural land.[7]

Relative costs of inputs

In recent years many governments of developing countries have favoured policies of subsidising the production and sales of agrochemicals to encourage food production. The total cost of subsidies can be very substantial, running into tens or hundreds of millions of dollars for some countries.[8] These subsidies have encouraged greater use of chemicals on the farm.

However, in Indonesia, the rice variety IR36, normally resistant to brown planthopper damage, actually suffered greater damage when treated too often with insecticides.[9] The more that farmers sprayed, the less likely it was that the natural enemies of pests would survive. It was in recognition of this kind of damage that the Indonesian government cut its subsidy of farmgate prices to half in 1985, and then to zero in 1989.

Agrochemicals may also be overused because of low cost relative to other inputs. In the USA farmers now pay about 25 per cent more for pesticides and 37 per cent more for fertilisers compared with 1977, but for labour the increased cost is 47 per cent, for tractors 75 per cent and for fuel 100 per cent.[10] Together, the low cost of chemicals and the high cost of labour discourage traditional methods of pest control, which are usually more labour- and time-consuming. Costs differ from country to country too: in 1980, for example, farmers in Sri Lanka, Bangladesh and Indonesia were paying less than US $0.35 for a kilogramme of nitrogen, yet in India, the Philippines and Thailand the cost was more than US $0.55.[11]

Farmers' perceptions

The views and beliefs of farmers are important – it is they after all who will enact any policies designed eventually to reduce agricultural pollution. Sometimes farmers incorrectly perceive both the pest problem and the means of control available. Farmers have a tendency to overestimate both the worse possible losses and the effectiveness of pesticides in reducing those losses.[12] It is also common for farmers to try to minimise further the costs of decision-making by treating all decisions according to a standard operating procedure, for instance by spraying at pre-set times without regard for the level or likelihood of pest attack.[13]

They may also hold particular views about the pesticides themselves. In the Philippines almost all rice farmers apply synthetic pesticides: they believe that pesticides are progressive and modern, and that powerful chemicals are a way of controlling one element of a hazardous environment, that produces floods, typhoons, and occasional droughts, as well as the pests.[14] In general they believe that higher frequencies of treatment are associated with higher yields, and if they could afford it would apply according to a calendar schedule.

But these beliefs do not accord with surveys conducted on farmers' fields, which suggest that at least half of sprayers do not benefit from increased yields. This mismanagement is to some extent rational. Their perceptions seem to have their beginnings in the extension activities during the early part of the Green Revolution: technical extension workers made repeated and aggressive demonstrations to persuade farmers to adopt insecticide-based methods of pest control. And yet when it comes to weeds, farmers are very capable of controlling weeds through a combination of cultural, mechanical and chemical methods, by which herbicides are effectively and appropriately applied.

More often sources of advice are informal, and standardised practices evolve in the place of the more finely-tuned and safe recommendations. In Mauritius, all vegetable growers stick to a seven-day harvest interval – the period between the last spray and date of harvest – because it suits the weekly harvesting pattern and spraying procedures.[15] Yet seven of the ten pesticide products in common use come with a safety recommendation that an interval of two to three weeks must elapse after the final spray before the harvest and consumption of the produce.

Preferences of customers

Another major cause of overuse is the need to protect high value crops, such as fruit and vegetables, when the appearance of the product to the consumer is of considerable importance. Such cosmetic control is particularly prevalent in industrialised countries. Supermarkets permit shoppers to select produce on the basis of appearance, and horticultural produce must sell itself or be left on the shelf.[16] High premiums are therefore paid for blemish-free crops: for example, wholesalers in the UK pay some 40 per cent more for top grade Cox's apples than for grade two apples.[17] Such premiums amply justify extra pesticide applications.

In the tropics there are equally strong pressures from wholesale customers of coffee and cotton cash crops for blemish-free crops. In Kenya the price for second-grade cotton is only half that for first grade, and low-grade coffee is sometimes rejected by factories.[18] These requirements make it impossible for farmers to grow these two crops without using pesticides.

Consumer preferences have also created demand for a whole new class of agricultural products, namely those grown in accordance with 'organic' practices. Consumers are willing to pay the higher prices for organically- grown products that cost more to produce. In industrialised countries the number of certified organic farms is growing. As a whole the public now perceives pesticides as high 'dread' phenomena, that is as presenting hazards regarded as uncontrollable, not easily reduced, involuntary, inequitable and of high risk to future generations.[19]

INTERLOCKING CHALLENGES FOR AGRICULTURAL DEVELOPMENT

I have detailed the potential stresses and shocks arising from productive agricultural activity that threaten environmental sustainability. But as Table 1 (p. 69) indicates, the costs are not borne by all equally. Moreover, those who suffer from the impact of pollution are not compensated for their losses. Agricultural pollution thus becomes a cause of reduced equity as well as sustainability.

However, the imperative to feed the growing numbers of people in the world remains. The central goal for development must be sustaining high levels of food production, ensuring that the basic needs of all are met, and minimising the damage and social costs arising from the agriculture producing the pollution. With these two goals in mind, people could conserve the natural resource base and ensure that its future use and exploitation would not be compromised. But evidence suggests that these goals usually conflict.

In the past twenty-five years managing the challenge of increasing food production has meant greater intensification, and by implication great changes in the natural and social environment. Take the structure of the countryside in the industrialised countries: removal of hedgerows, woodlands and ponds has reduced the costs to the individual farmer by cutting the number of times machines need to turn in the now larger fields, reducing hedge maintenance costs and producing small gains in the area of cultivable land. But the largely hidden internal and external costs are also significant. Wild patches have aesthetic value, they harbour predators of crop pests (but also some pests) and other wildlife, they provide shelter for livestock, and in the lee of trees and hedges windspeed is reduced, which keeps down wind-induced erosion and evaporation losses, and increases humidity, temperature and soil moisture content.

In the intensive Green Revolution regions in the developing countries there have also been very significant changes in social structure. The modern varieties of cereals tended to be adopted only by those farmers with the resources to purchase the inputs necessary to achieve the yield increases. Moreover, the labour required declined as labour-intensive activities were replaced by alternatives, such as machinery, chemicals or new practices.

This has, in turn, led to changes in the relations between men and women. In Indonesia, for example, teams of women formerly used traditional *ani-ani* knives to harvest the rice, and in part-payment received a proportion of the rice. But with modern varieties of rice all coming to maturity together, standing crops were increasingly sold to merchants who hired specialist male labourers to harvest with sickles. The result was greater efficiency, but also the isolation of women from the process of production.

INTEGRATED APPROACHES FOR POLLUTION CONTROL

The matching of these apparently conflicting goals in agricultural development can only be achieved through the extension and adoption of integrated approaches to agricultural management, such as technologies encompassed by agroforestry, integrated pest management, integrated nutrient conservation, and soil and water conservation. There are now a wide range of known technologies for all of these integrated strategies that will enhance yields and reduce environmental contamination and which, put in economic terms, have the potential to increase private returns whilst reducing both private and social costs.

The possession of this knowledge is not enough. All these environmentally sensitive technologies are subject to the same problems as any intervention designed in the research station or government department. The planning for management is done without involving local people in the process, and consequently may fail to reach desired levels of adoption.

For example, alley cropping of leguminous trees amongst cereal crops, shown to be highly productive and environmentally sensitive on research stations in West Africa, has been adopted by only small numbers of farmers. Many integrated pest management programmes have failed to be economically viable to farmers despite proven scientific success.

A central feature of the pollution so far described is the widespread nature of the causes and impacts. It cuts across administrative boundaries, presenting severe problems for single-focus regulatory agencies. Lack of liaison and coordination of activities has, as a consequence, meant that all aspects of the environment are rarely considered together.

The key to successful natural resource management for sustainable agriculture lies in a partnership between the research scientists, policy makers, regulators, developers and extension workers plus the farmers and rural people themselves. There are two important reasons for this.

First, each of these integrated approaches requires a greater range of scientific knowledge and understanding if all the apparently conflicting goals are to be met. Second, each of these integrated approaches requires detailed local ecological and socio-economic information on livelihood systems – who better to provide this than local people themselves. There is nevertheless some evidence of success.

In the Sacramento Valley of California, controversy recently erupted over the two herbicides thiobencarb and molinate, used on irrigated rice.[20] Following their escape from fields in irrigation water they appeared in drinking water, producing a bitter and unpleasant taste. To prevent regulatory action the rice industry had to defend the use of herbicides in the public domain. An educational programme was developed by the rice industry, the local water quality control board and local researchers encouraging growers to adopt 'Best Management Plans' to reduce the pollution. These plans recommended holding the irrigation water in the fields for a set period after application, and the recycling of irrigation water. These practices were rapidly adopted and, as a result, concentrations in drinking water rapidly fell.

In the UK, the first report of the recently established integrated pollution control agency, Her Majesty's Inspectorate of Pollution, has already drawn attention to some of the advantages of integrated pollution control.[21] Recognising that precautionary measures should be taken where the evidence is not yet conclusive, their early studies also were important in demonstrating how inspectors with different skills and from different disciplines could work constructively together.

In the uplands of the Philippines spontaneous adoption of different types of contour hedgerows is now taking place, as a result of the innovations now being jointly developed by the farmers and researchers.[22] Elsewhere projects and programmes in Africa, Asia and Latin America are beginning to take the required steps to support this partnership for sustainable, productive and equitable agricultural development.

TABLE 6.1 *Key routes through which agriculture pollutes the environment, and range of sectors suffering these imposed costs external to agriculture.*

Pollutants	*Consequences*	*Sectors on which costs imposed*
A: Contamination of water		
Fertilisers: nitrates	Methaemoglobinaemia in infants; cancers	Water consumers, Health services
Fertilisers: nitrates and phosphates	Algal blooms causing taste problems; surface water blockage, fish kills; coral reef destruction; illnesses and deaths due to algal toxins	Water treatment works, Surface water transport, Fisheries, Tourism, Water consumers, Agriculture
Organic livestock manures: nitrates	Algal blooms, plus deoxygenation of water	as above
Soil	Physical blocking of surface water courses; filling of lakes, reservoirs and harbours; disrupting irrigation flow in channels; destroying fisheries and coral reefs	Fisheries, Navigation, Agriculture, Water Supply, Tourism
Pesticides	Contamination of rain-fall, surface and ground water, causing wildlife deaths and exceeding standards for drinking water	Water consumers and suppliers, Agriculture, Wildlife, Health services

B. Contamination of food

Pesticides	Pesticide residues	Food consumers

C: Contamination of air

Livestock manures	Ammonia, playing role in acid rain production; odour nuisance	Agriculture, Industry, Wildlife, Trees, Buildings, Aesthetic value
Fertilisers	Nitrous oxide, playing role in ozone depletion and global warning	Costs occuring in future (relating to health, agricultural disruption, sea-level rise)
Paddy rice	Methane, playing role in global warming; and ammonia, playing role in acid rain	as above
Ruminant livestock	Methane	as above
Biomass burning (of forests, of cereal straw)	Methane, nitrous oxide, plus other combustion gases, smoke and particulates	as above

7 Halocarbons and Stratospheric Ozone — a Warning from Antarctica

Joe Farman

The depletion of the ozone layer has been the outcome of an avenue of development along which society had been hurrying so rapidly that it had overlooked indications of a 'Dead End' around the corner. It seems that by good fortune rather than good judgement, an 'End' may now be in sight, with no one dead or harmed. The agreement reached by ministers in Brussels in March of 1989, accepting the need to phase out completely the production of those CFCs and halons listed in the Montreal Protocol, shows that the seriousness of the situation has at last been recognised. It is encouraging, since in many ways the threat to the ozone layer can be regarded as a test case. If the world lacked the will to tackle what is essentially a simple technical problem, there would be little hope of its solving the more complex environmental and development problems which we are facing.

In looking at the issue of the ozone layer, I should like to begin by establishing what stratospheric ozone is, what it does, and how its depletion was discovered. I shall then outline the chemistry involved, look at the damaging chemicals and their uses in society, and point out a few guidelines for the way forward.

The 'ozone layer' which has been the object of so much concern recently is situated in the upper atmosphere, a region called the stratosphere. The depletion of it is cause for concern because ozone is the only constituent in the atmosphere that is present in sufficient quantity to stop the very short wave, ultraviolet (UV) radiation of the sun from reaching the ground. It is very important that this radiation does not reach the ground since it can adversely affect the genetic material of all living organisms, including man, and a wide range of materials. However, the ozone layer is not only important because it protects living organisms at the earth's surface, but also because in absorbing the radiation the ozone layer generates heat in the upper atmosphere. This warm region has similar temperatures to the earth's surface, and as a result controls the whole temperature structure and circulation of the middle atmosphere. If ozone continues to be depleted, not only will the effects of increased UV radiation be felt, but also the temperature structure of the middle atmosphere could be upset, the effects of which we are completely unable to predict at the moment.

Ozone is present in the stratosphere in an equilibrium of destruction and creation. It is created by a photochemical reaction between oxygen and sunlight. It is destroyed in chemical cycles involving constituents present in very small quantities. It is halogens (fluorine, chlorine, bromine and iodine), but particularly chlorine and bromine that are the most worrying destroyers of ozone. It was the realisation that these destructive chemicals need only be present in trace proportions to cause damage, that led to an awareness that the amount of ozone in the atmosphere could be significantly altered by human activities. There is now no doubt that this has occurred.

In the last decade, a new feature has appeared in the annual variation of ozone over Antarctica. This feature is a deep minimum in spring (September–October). Satellite maps show that the most severe depletion – the so-called 'ozone hole' – occurs in the very cold air trapped in the polar stratospheric vortex (meteorological conditions peculiar to the poles). In 1987, the depletion was the most severe yet seen. Between August 15 and October 13, the heart of the ozone layer disappeared over the British Antarctic Survey's monitoring station at Halley Bay, leaving only the wings of the earlier distribution. Between 14.5 and 18.5 km altitude, 97 per cent of the ozone was destroyed. It can be seen from this that the chemistry within the polar stratospheric vortex, which is effectively an isolated air mass, has been severely perturbed. The removal of ozone has reduced the heating of the lower stratosphere, and low temperatures persist in the core of the vortex for several weeks longer than previously. The temperature at 14.5 km altitude over Halley Bay on 1 December 1987 was – 69°C, some 15 degrees below the extreme value recorded between May 1957 and April 1975. As I have said above, the possible consequences of these temperature changes have yet to be explored but are nonetheless worrying.

This seasonal depletion of ozone in Antarctica, the first unequivocal systematic change in ozone climatology to be identified, was completely unexpected, and has yet to be completely explained. However, results from the American expeditions in 1986 and 1987, and from the Airborne Antarctic Ozone Experiment in 1987 leave no doubt that the main cause of the depletion has been the growth of inorganic chlorine in the stratosphere. This growth of chlorine has arisen almost entirely from the emission of man-made chloro and chlorofluorocarbons (CFCs).

Although chlorine is the major agent for the depletion of ozone, on a per atom basis bromine is a more effective one. Measurements in the 'ozone hole' in September 1987 showed that the abundance of the bromine monoxide radical[1] was above 1 per cent of that of the chlorine monoxide radical.[2] It is estimated that the bromine monoxide radical has accounted, nevertheless, for between 5 and 10 per cent of the ozone destruction. I shall now attempt to explain why it is that these chemicals get into the stratosphere where they can destroy ozone, but to do so I must first establish their nature.

A halocarbon is any substance which can be produced from hydrocarbons (substances composed solely of carbon and hydrogen) by replacing one or more of the hydrogen atoms by halogen atoms (fluorine, chlorine, bromine or iodine). The chlorofluorocarbons (CFCs), that have become so infamous, are halocarbons that contain both chlorine and fluorine, but no other halogen. The relevance of these halocarbons is that if they are broken down in the stratosphere the halogen atoms that are released destroy the ozone layer. An important distinction to be made is the difference between halocarbons that are *fully halogenated* and those that are not. When all the hydrogen atoms on a halocarbon have been replaced by halogen atoms, the halocarbon is said to be *fully halogenated.* On the other hand, if there are any hydrogen atoms remaining, it is said to be *not fully halogenated.*

Fully halogenated halocarbons can only start to be broken down by UV light. This is a very slow process below the ozone layer and in practice it takes place above it when almost all of the halogen atoms carried by such halocarbons are released into the ozone in the stratosphere. If the halocarbon contains hydrogen atoms, there is an alternative reaction available with a hydroxyl radical,[3] present in atmospheric water vapour, and this is usually the faster process. The breaking-down of the not fully halogenated halocarbon then occurs largely in the lower atmosphere. The halogens released there are converted into inorganic compounds (mainly acid halides, such as hydrochloric acid), which, being very soluble in water, are rapidly returned to the earth's surface in rain. Thus, most of the halogen atoms carried by this class of halocarbon never reach the ozone layer. Natural or industrial releases of inorganic halogen compounds from the earth's surface are similarly of little significance to stratospheric chemistry, except possibly when associated with large volcanic eruptions, whose plumes occasionally penetrate the stratosphere, or with very large explosions.

The 'lifetime' of a gas is an important concept. If further release of a gas could be prevented, and the destruction rate maintained, the gas would be removed completely after one 'lifetime'. In practice, the destruction rate is proportional to the quantity of the gas, which decays exponentially, falling after one lifetime to 37 per cent of its value at the time of cessation of release. Consistent results are now being obtained for the lifetimes of some of the more abundant halocarbons, but there are many unresolved problems. Nevertheless, 'lifetimes' of gases are good indications of their destructive potential: the longer the 'lifetime', the more destructive the gas can be.

The abundance of chlorine in the atmosphere has risen dramatically in this century. Until 1900, the primary source of stratospheric chlorine was methyl chloride,[4] a halocarbon with large natural sources and a short lifetime, about 1.5 years. The atmosphere now contains four man-made halocarbons each carrying amounts of chlorine comparable to that in methyl chloride. Two other man-made halocarbons are accumulating rapidly. Four of these six are fully halogenated. They are therefore of the more damaging variety of halocarbon, and have lifetimes of up to 140 years. The oldest is carbon tetrachloride:[5] first produced in the early 1900s, it was used as a solvent and as a fumigant, and later to produce CFCs.

The CFC industry began in 1931. Sustained exponential increase in releases of F-11[6] and F-12[7] followed up to 1974. Releases then declined slowly until 1982, for various reasons, including the ban on CFCs in aerosol products in the United States and a few other countries, and uncertainty over the introduction of regulation. However, releases increased again from 1982 to 1986. Release of methyl chloroform,[8] used largely in solvent and adhesive applications, has stabilised somewhat after a period of very rapid growth. F-113[9] release has increased dramatically in recent years. It is used as a solvent in the electronics industry. Release of F-22[10] also appears to be rising sharply.

Fluorine plays little part in ozone depletion. The halogen, when released from CFCs, is rapidly converted to hydrogen fluoride. This halide is practically inert in the stratosphere, and is rapidly washed out of the troposphere. No natural source of atmospheric fluorine has yet been identified. At present the main carrier is F-12. Interestingly, the next most important carrier is carbon tetrafluoride.[11] This is a very inert gas, with a lifetime exceeding 10,000 years. Essentially emission and accumulation rates are the same. Given that the carbon-fluorine bonds have very powerful 'greenhouse effect' qualities, this is cause for concern, even if the fluorine is not likely to damage the ozone severely.

Atmospheric bromine was until recently primarily carried by compounds which are not fully halogenated. Typical of these is methyl bromide,[12] which has both natural and industrial sources. It has been used as a grain fumigant, and is emitted from the exhausts of cars using leaded petrol. Both uses are probably declining. Its lifetime is short, about 1.7 years. It is now thought that very little of the bromine carried by these compounds reached the stratosphere. Three long-lived fully halogenated bromine carriers are now being released in small, but apparently rapidly increasing, amounts. Halon 1211,[13] with a lifetime of about thirty years, is used in portable fire-extinguishers. Halon 1301[14] is used as a fire-extinguisher in total-flooding applications where carbon dioxide or water are not acceptable (in computer rooms, engine rooms, museums etc). Halon 2402[15] is reportedly used in Eastern bloc countries in similar applications, but few details are available. These halons are probably already the major carriers of bromine to the stratosphere. Bromine compounds are now recognised to be very efficient catalysts for ozone destruction. Tolerable concentrations are much less, by a factor of at least thirty, than for chlorine compounds. The atmospheric concentrations and industrial production of these halons should be monitored closely.

I should now like to highlight a few points of interest that indicate where progress needs to be made. Everyone in the West has said that they have stopped making carbon tetrachloride,[16] and indeed it has been banned, yet it is still growing in quantity, with no known natural sources. F-113[17] is used mainly in the electronics industry. The world's largest market for electronics is the United States military, who insist on using cleaning agents like F-113, even though it may now be the case that high-pressure water will suffice. Moreover, the claim that natural resins must be used seems curious when there are now very high quality synthetic fluxes available which do not require such cleaning agents. Carbon tetrafluoride,[18] the second highest carrier of fluorine, is an incidental industrial release, mainly from aluminium smelters. The release rate is inferred to be some 18 kilotonnes per year. Direct, intentional, production by the CFC industry has never exceeded 60 tonnes per year. The accumulation of this halocarbon is a striking example of a practically irreversible change in atmospheric composition caused, inadvertently, by man's activities. Manufacturing figures for

Halon 1301,[19] the compound used as a fire extinguisher in total flooding applications, suggest that of all the releases of this gas, only 6 per cent have been used to put out fires; the rest have been released for tests on behalf of insurance companies and the like, or as waste.

The accumulation of the halocarbons themselves has made a significant contribution, since 1970, to the radiative forcing of climatic change, the so-called 'greenhouse effect'. This is a direct effect of the halocarbons on climate, and is distinct from, and additional to, climatic consequences of depletion of the ozone layer. The continued release of long-lived halocarbons can no longer be tolerated. Safe substitutes must be found, and large-scale production of substances which can carry chlorine or bromine to the stratosphere should cease. Only when this has been achieved will the atmosphere begin to return to a safer state.

Part Three: Individuals, Society and Sustainable Development

8 Changes in Perception

Martin Holdgate

A DIVERSITY OF STARTING POINTS

The world is a highly diverse place, and individual perceptions of the environment clearly depend on where people live, on their economic circumstances, and on their cultural and religious backgrounds. The Bedu of the desert, Inuit of the Arctic, Alacaluf of the Western Patagonian Channels, and San of the Kalahari interpret their environment in the light of thousands of years of experience and adaptation. The forester, hunter, fisherman, farmer, urban labourer, industrial tycoon and international civil servant each has his own special viewpoint. Moreover, perceptions of the environment inevitably change with experience, and the rates of such change also differ widely. Global generalisations about the perceptions of the environment today or a decade hence are impossible.

We can, however, be clear about one thing. Such diverse individual perceptions are the essential foundation for the actions that must be taken to care for the world environment. For the earth is moulded by the behaviour of human individuals. It is a convenient shorthand to demand action by governments, as regulators of the many components of complex modern societies, but those governments can only act effectively if they carry their people with them, and people are notoriously obstinate when their view of what action is needed differs from that of their rulers. The key to the future is to know how to inform people and develop their understanding of the environment so that they willingly take the actions that are required. This information has to be tuned to the varied starting-points of many diverse individuals.

The cultural and linguistic dimension must not be ignored. Language is not just a means of communication between minds with identical thought processes. The thought processes themselves, and the way in which they convey complex imagery are intimately interrelated to the language used as the medium of communication, and it is necessary to be sensitive to this in understanding perceptions and forging mutual understanding. This point was well brought out in a seminar held at Fontainebleau in October 1988 to celebrate the 40th Anniversary of the founding of the International Union for Conservation of Nature and Natural Resources (IUCN). In his summary, the Chairman commented that 'different cultures have a definite contribution to make, because different cultures have held, and still do hold, different views on nature and on the relationships between man and nature. In point of fact, the human race is constantly pursuing complex goals. If these different goals – and the protection of nature is one – are to be taken seriously, due thought must be given to the question of arbitration and the ways and means of achieving compromises.'[1] These are words that we should bear in mind as we evaluate popular perceptions of the environment and look to the future.

A HISTORY OF DISLOCATION

Nearly all our ancestors were country-dwellers. They lived from, and were always close to, nature and they had an intense awareness of that interdependence. The early sacred writings of many religions reflect this situation. Folk customs, like those reviewed by Sir James Fraser in his classical volumes *The Golden Bough*[2] record many formulations of reverence for, and care of nature. The sacred grove of trees is more ancient than the temple. The recognition that sacred life abides in trees and animals pervaded traditions in Europe, the Americas, Africa, Australia and Asia. Those who have carried traditional ways of life through into modern times have often expressed the relationship eloquently, as Chief Seattle did in his letter to the American President in 1885: 'Every part of this earth is sacred to my people. Every shining pine needle, every sand shore, every mist in the dark woods, every clearing and humming insect is holy in the memory and experience of my people.'[3]

Over the centuries many such cultures had evolved ways of life that were in harmony with nature. Certain nomads are said to have tied a bag of grass seeds about the neck of the lead animal of a flock, with a slit that released a trickle of seeds as the flock moved, so that the seeds would be trampled into the ground and fertilised by the dung of the animals that followed. Cultivation procedures which achieved a balanced relationship with nature included the early swidden (or shifting) system in forest land, which is stable in areas of low human density, where the areas of abandoned cultivation revert quickly to forest, and where the forest itself remains untouched for many decades before a particular area of land is likely to be cleared again.[4] More elaborate systems including modifications of water flow are also ancient. The irrigated gardens in the highlands of Papua New Guinea date back some 5,500 years,[5] and the irrigation systems in Mesopotamia were the foundation of civilisation over 7,000 years ago.[6] Such systems were commonly governed by rules for the management of shared resources, like those governing the abstraction of water from the *falaj* channels of Oman, where water allocations are based on time units and on a complex set of social relationships.[7] No doubt similar rules governed the management of the great irrigation systems on which the civilisations of the Tigris and Euphrates Valley depended many millenia ago. Similar social systems for apportioning shared resources can be seen even today amidst the agriculture of modern Europe. For example, common grazings in parts of northern England continue to be regulated according to a system of 'stints' which rates the capacity of the grazing in terms of numbers of horses, cattle or sheep, and give certain properties in the village a specified number of stints as an integral component of their holding.

It is a matter of common observation that however balanced these kinds of human uses were with the environment in times past, the history of humanity has been one of change and breakdown in such relationships. I suggest that there have been three major factors behind such breakdowns.

First, as human population density has increased, the maintenance of the traditional systems has become more difficult. Swidden cultivation breaks down when the period during which the land is under forest becomes too short for fertility to be fully restored, or for the true forest ecosystem to re-establish itself. Grazing management systems break down when herds become too big, and their capacity to find areas of vegetation flushed by rain becomes too effective (as in Saudi Arabia today where motor trucks transport animals to areas announced on the radio as having had recent rain). The irrigation systems of Oman are at risk because increasing numbers of people are sinking wells from which they are pumping water at a rate exceeding natural replenishment. The water tables drop, the *falaj* systems are undermined, and date palm groves die because their roots can no longer reach the water table.

Salinity, always a problem in irrigated arid lands, is said to have destroyed the Mesopotamian cultivation systems in ancient times, possibly during a period when the social structure was under threat from war. It is somewhat ironic that recent reviews have emphasised as a key to the sustainable development of arid lands the restoration of tight systems of management of shared water supplies![8]

A second factor has been the arrival on the scene of new and invisible human pressures on nature. Air pollution is an obvious example. Acid rain and ozone depletion in the stratosphere are invisible to those suffering from them. Pollution has created changes that are both unexpected and inexplicable even by sophisticated societies like those living among the forests and lakes of western Europe or in North America and Japan. Rural communities, however great their sensitivity to nature, cannot accommodate these kinds of pressures since they are caused by other groups at great distance. The causative agents have to be traced to source, and those responsible have to be persuaded to take corrective action – and this requires communication at government level. Meanwhile, the most that affected rural communities can do is adapt, protest, or accept their losses and diminished quality of life.

A third cause of the breakdown is simply the scale of the disruption. Many of the disturbances to environmental systems today are continental or truly global in extent. The 'greenhouse effect' threatening the equilibrium of global climate and the level of the sea, the depletion of the stratospheric ozone screen, and trans-continental acidification can only be corrected by collective international effort. Action has to be taken by the governments of sovereign states other than those experiencing the damage, which perhaps have other priorities for the use of their national resources. Certainly the scale of the problem depersonalises it. It is perceived as something done by 'them', far away., and outside the responsibility of the individual. And the perceived ability of the individual to do anything useful to solve so large a problem is tiny. The obvious inconveniences of not using fossil fuels or chlorofluorocarbons outweighs the hypothetical value of that person's contribution to environmental protection.

In the developing world today, many changes, especially those resulting from mounting human numbers, are undermining old perceptions and balances between humanity and nature without creating new ones. What has been aptly called 'the pollution of poverty' is rife in rural areas, especially in the many countries with rates of population increase varying between 2.5 and 4 per cent per annum. Such pressures lead to the cultivation of soils unsuited to intensive use and unable to sustain it, inefficient irrigation systems leading inexorably to salinity, and the deforestation of slopes, which then become prone to erosion, and of upland catchments which no longer regulate and maintain a flow of water through the dry season. The people undertaking these acts generally know that what they are doing is unwise and at variance with social tradition, but they are impelled in this way by the overriding necessity to live tomorrow.

Another disruption, also linked to the surge in rural populations and the accentuation of rural poverty, is the vast migration of people from the land to the cities that many countries have experienced. Such migrations detach people from their roots. In the urban areas the man-made environment commonly offers squalid shelter, unhealthy air, impure water and a threat of disease, but in compensation there is the prospect of employment and supporting services which are better than in the rural hinterlands. Such migration also places the urban dwellers in an environment over which they have little influence. They can do almost nothing about air or water quality, except tolerate them.

In developed countries, too, urbanisation has inevitably meant the detachment of human communities from the land. As in the developing countries, certain environmental needs like clean air and pure water cannot be guaranteed by individual efforts, and much of the damaging pollution of these media is not visible, and hence cannot easily be reacted to. Less than a century ago, many urban populations in the early industrial centres of Europe and North America depended for their livelihood on industries that externalised costs and enhanced their profits by discharging or dumping fumes, liquid pollutants and toxic solid wastes. There was little social condemnation of such pollution except from a few very perceptive individuals. The wealthy industrialists moved out to live in cleaner and greener suburbs: the work forces put up with their lot for the sake of employment, and while there were campaigns to improve the environment they had to fight every inch of the way against the forces of commercial profit.[9] Very possibly, the long-term threat posed to health and well-being by these conditions was simply not a matter of concern, when judged against the benefits of immediate employment. It is after all human nature to rank pleasure today more highly than sorrow deferred, as the continued prevalence of tobacco smoking shows.

The developed countries today have what are commonly called 'consumer societies', perhaps more accurately described as 'producer societies'. Employment and individual wealth depend on the manufacture, sale, and servicing of products. High consumption of such products is equated with a high standard of living, and indeed forms the basis of measurements of Gross National Product, and thus the place of a country in the international wealth table. The fact that producer societies often squander environmental resources by their emphasis on short-life, quick-discard products is not taken into account in the equations. Only a few outspoken prophets, regarded by many as cranks and by the champions of production as pernicious, attack the advertising which persuades people to replace the things they now have by hypothetically better 'new models' or to consume more than is needed to give comfortable enjoyment of life. Yet the measurement of prosperity and success in terms of goods consumed, and the competitive nature of such assessments, are clearly at a variance with any rational judgement of the need for a sustainable use of the resources of the biosphere.

One of the great dangers of today's world is that the very perceptions and goals which have caused the developed countries to consume a disproportionate share of the world's environmental resources, and to lock themselves into patterns of economic activity which are unlikely to be sustainable, are themselves being transferred, as goals to emulate, to the developing countries. Unless there is a collective shift in perception of what the world community needs to do to achieve a stable balance between humanity and nature, competition for increasingly inadequate resources is likely to lead to disruption and even strife. National frontiers may no longer be respected when there are clear contrasts in the human well-being on either side of them. There are already movements of people across some such frontiers (for example between Mexico and the United States and the two Germanies) and many analysts of the world's environmental future fear that much larger movements, and even wars, may be sparked by environmental inequality. Of course nature always regulates populations, but the methods employed do not always square with what we are pleased to call civilised values.

THE GROWTH OF AWARENESS IN THE DEVELOPED WORLD

Environmental awareness, in the wide sense, is the accumulation of all the myriad perceptions of their surroundings among the people of the world – people in the midst of the many dislocations just described. But the term is commonly used in a more restricted way. Statements about 'the growth of environmental awareness' generally describe a relatively limited phenomenon, centred in the developed countries of Europe,

North America and Japan, and especially among the more prosperous sectors of those societies (including government leaders). Many reviews also focus on the manifestations of this awareness within anglophone cultures. There is, therefore, a real danger of generalising from a biased position.

Within that cultural and geographical context, however, environmental protection movements now have a respectable history reaching back over a century. In both Western Europe and North America they began with the establishment of voluntary societies (non-governmental organisations) dedicated to saving outstanding landscapes, historical buildings and monuments, and protecting birds. The growth of 'The Environment' as a concept, and the use of 'Environment' as a term of everyday life is of recent origins, certainly dating from the years after the 1939–45 war, and very largely a phenomenon of the period since 1950. The movement undoubtedly acquired extra impetus in the late 1960s and early 1970s, especially at the time of the United Nations Conference on the Human Environment, held in Stockholm in June 1972. The development of awareness and concern, it must be added, has often proceeded with only the haziest idea of what the word 'environment' means and with its interpretation in different ways by different sectors of the community. For some 'environment' is seen relatively narrowly in terms of National Parks, protected areas and the habitats of wild animals. For others it is essentially the natural world, but not the city, the motorway or the factory. For others it is perceived broadly, in terms of the definition attributed to Einstein: 'environment is everything that isn't me!'

This lack of consensus on definition is really unimportant. What is important is that people have become increasingly aware of the dependence of human societies on the resources of the planet and the continuing maintenance of the essential processes of the biosphere. How has this come about?

There have been three contributing sources: formal education, scientific research and publicity through the mass media. Formal education, at least in Europe, appears to have followed rather than led. Universal education in elementary science is in itself a recent phenomenon, considered scarcely necessary or respectable for some sectors of the community (especially females). The sort of physics, chemistry and biology taught up to thirty years ago was of little relevance to the understanding of the natural world. Social and economic studies were rarely blended with natural science to give insight into the workings of the system of interaction between people and nature. As a whole, formal education has played little part in awakening people to the environmental problems of the modern world and to the need to treat it differently.

Science did not make a major contribution either. Ecology evolved as a specialist discipline, largely concerned with elucidating the working of natural ecosystems. Human ecology, generally the province of geographers, was rarely regarded as a branch of the same discipline as 'scientific ecology', and the outputs from the latter tended to take the form of highly technical and specialised publications inaccessible to (and, by their nature, of little interest to) the ordinary citizen. Only a few visionary and committed scientists (like for example, the late Rachel Carson) took upon themselves the role of the prophet and explained modern scientific understanding of the natural world in terms that the ordinary citizen could understand.

As a result, modern environmental awareness has largely been created by journalists, radio commentators and television presenters. It has inevitably been built on 'news stories' which touch the emotions of those they address. The vast majority of individuals alerted by those stories have neither experience of nor involvement in the subject matter. The stories are commonly made memorable by catchy headlines like 'Black Tide of Death', 'Have We Poisoned the Sea?' or 'The End of the Forests'. Some stories are fair reflections of genuine underlying fact: others exaggerate or trivialise. They have all, however, performed the important task of getting the message across to the literate community that the world is in a mess and that people are responsible.

The scale of this publicity has fluctuated. Before the 1960s, it was largely a reflection of widely separated 'trigger events', The deaths of more than 4000 people in the disastrous smog of London in 1952/53 was one such event. The similar incident in New York in 1963, causing 800 deaths, was another. It is no accident that an analysis of articles in the *New York Times* dealing with environmental topics in 1960 focused on air pollution and water pollution, made little mention of environment as a general theme, and incidentally numbered well under 150 articles all told.

During the 1960s and 1970s the surge of concern meant that environmental articles were more regular, and that the emergence of specialist environmental correspondents began to guarantee both their consistency and professional accuracy. The scientific community contributed to this informed comment through books like Rachel Carson's pioneering *Silent Spring*[10] and other works like those by Paul and Ann Ehrlich[11] and Barry Commoner[12]. Authoritative reports on environmental issues also began to appear. In the United States, for example, the President's Science Advisory Committee reported in 1967 on *Restoring the Quality of our Environment*, and a special committee on the study of critical environmental problems reported in 1970 on *Man's Impact on the Global Environment*. As 1972 and the Stockholm Conference approached, there was an explosion in the number of environmental articles. The *New York Times* in 1970, for example, had some 150 general articles on environment, some 650 articles on air pollution and some 900 articles dealing with water pollution.

That wave passed, and by 1979 the *New York Times* was down to under 50 environmental articles, under 250 on air pollution and under 400 on water pollution. Trigger events like the Chernobyl disaster, the discovery of the ozone hole over Antarctica, the oil spill at Valdez in Alaska, and elephant poaching in Africa have continued to provoke surges in media coverage, but it was only in the mid- to late 1980s that the consistency and generality of concern manifested itself in a second explosion in thoughtful articles covering the whole environment and the total human predicament. Although major wide-circulation magazines like *Time* and the *National Geographic* have carried environmental articles for many years, a new phenomenon has been their devotion of entire issues to environmental concerns.

There is a danger in the way in which environmental issues have been reported in the mass media of the developed world. They have to a considerable degree externalised individual responsibility. People are not generally involved in the story except as victims or (as in the case of oil spill cleanup) as workers struggling to restore a devastated ecosystem. Few people have direct personal experience of the incidents reported. The result is that the tendency is to attach blame to, and demand action from, somebody else. The responses individuals have often been called upon to display have been channelled through intermediaries. Environmental organisations share the responsibility for this, for their average news story can often be summarised, in general terms, as 'stop them – fund us'. Very few media campaigns are designed to persuade individual people to curb their own bad habits. This is perhaps partly because such action would contradict the central thrust of the producer/consumer society. People want to have their cake and to eat it, and it takes a courageous media campaign to urge people not to buy the kinds of products which the newspapers carrying the campaign might themselves be advertising. It may also be due to the fundamental point, made previously, that one person's contribution to the overall good is often so small as to be imperceptible. Although campaigns

against smoking, aerosols using ozone depleting chemicals, litter, an
consumer goods judged harmful to the environment have bee
mounted, and family planning has been urged increasingly on environ
mental grounds, these have been largely impelled by committed group
and organisations propounding a code of values that makes people fee
that their personal contributions are truly worthwhile.

Such groups – among them Friends of the Earth, Greenpeace, and th
Worldwide Fund for Nature – have gained greatly in strength, as hav
more traditional conservation bodies, like the National Trust or th
Royal Society for Protection of Birds in Britain, the National Audubo
Society, Wilderness Society and Sierra Club in the USA, the Moscov
Society of Nature Investigators in the USSR, and the numerous societie
in France, Switzerland, the Nordic countries, the Netherlands, th
Federal Republic of Germany, Canada and other parts of the develope
world.

They have also moved into non-formal education, promoting courses
distributing literature, and establishing information centres suitabl
for the use of school groups and youth clubs. In parallel, the forma
education establishment in a number of countries has begun t
take a much keener interest in environment, often impelled by teachers
concerns aroused by the media.[13] A major international Conferenc
on Environmental Education held at Tbilisi in 1977 made numerou
recommendations including a recognition of the role of the mas
media, and formal consumer education has become included in man
home economics courses. In Japan, a citizens' movement agains
pollution, which also involved education designed to preven
environmental disruption, began in the 1960s. In 1971 the Japanes
Parliament requested that this *Kogai* education be introduced t
schools, and since that date it has become an integral part of education a
all levels.[14]

Industry, too, has recognised that a more responsible attitude toward
the environment is important for its own future. Under the leadership o
the United Nations Environment Programme and the Internationa
Chamber of Commerce, a World Industry Conference on Environ
mental Management was held at Versailles in 1984.[15] Its aim was th
development of guidelines on how to combine economic growth wit
sound environmental management. Since then industrial groups i
many countries have had their own conferences and campaigns, such a
the Industry and Environment Year mounted in the United Kingdom b
the Royal Society of Arts. Awards to industry for inventions an
developments that are of particular environmental promise have als

been instituted in various countries, including the Pollution Abatement Technology Awards Scheme in the United Kingdom, later converted to a European Better Environment Awards For Industry Scheme in the European Year of the Environment.[16] Industry has given increasing emphasis to the environmental dimension of its work, giving publicity to environmentally benign products, and developing recycling of materials such as glass, metals and paper in many developed countries.

Government responses have tended to follow rather than lead popular demand. But the swelling concern for the environment in the 1970s was reflected in the establishment of an increasing number of Departments of the Environment in many countries, and by a burgeoning in environmental law and official publications, including reports on the state of the national environment (published in the USA, UK, Japan, India, and several other countries). Such Departments have, however, been implanted as sectors alongside traditional Departments of Commerce, Industry, Finance, Agriculture, Energy or Transport, thus denying the horizontal nature of environment and the fact that its resources sustain all sectors of society. The World Commission on Environment and Development has emphasised the need for an inter-sectoral approach, but governments have been relatively slow to respond – an exception being provided by Norway, which has created a mechanism, led by the Minister for Environment, under which the annual spending plans of all Departments of State are scrutinised from an environmental standpoint and something of an 'environmental budget' is prepared, published and discussed by Parliament. Public demand has not yet increased to the point at which a radical reconstruction of government to accommodate universal environmental concerns has become unavoidable.

Most people in the developed countries have yet to recognise their own personal involvement in global systems of interdependence. Although they consume a large part of the world's natural resources, they do not recognise their impact on the Third World. For example, until it was highlighted by recent media campaigns, few knew about the crippling burden of debt which the poorest countries have incurred, or that this was in part a consequence of loans for schemes which were intended to support their economic development, but which failed because the schemes adopted were ill-attuned to the sensitivities of environment and culture. The dominance of world trading patterns, world prices and world investments by the consumer/producer societies of the northern hemisphere remains great. It is demand in the 'North' that drives the export-orientated economies of the 'South', but curbing that demand is not the answer, and could indeed aggravate th 'pollution of poverty' there. What is needed is a willingness o individuals in the North to pay more for the goods from the South and for economic interactions to be adjusted to support genuinel sustainable development there.

THE GROWTH OF AWARENESS IN THE DEVELOPING WORLD

Environment perception in the developing world is also changing. It ha however, been articulated far less, and in most countries features les prominently in the media than in the developed countries. However, th developing world still displays a much greater prevalence of tradition understanding of the relationship between communities and enviror ment. Many more communities continue to live in close relationshi with the natural world, and there is less consumer aggression. Man projects of sustainable development are being carried forward, designe to cultivate or maintain harmony between local agricultural an pastoral communities and the environment around them.

The key to the success of these projects is that they are rooted in th way of life of the communities undertaking them. I shall give on illustration of this. In the eastern Usambara mountains of Tanzania with support from the European Community, and under the overal supervision of a Tanzanian project leader, IUCN has appointed villag coordinators in thirteen communities. These coordinators are youn people from the communities themselves, where they work to develop various activities to increase local self-sufficiency and to enhance the sal of products from the villages, bringing returns directly to thos communities. Villagers are involved in these activities at all levels Schoolchildren survey contour lines, some individuals run nurser gardens for vegetables, spice plants and others for planting alon contours to stop erosion. Other people raise tree seedlings to be plante for fuel wood. The whole community helps to build fish ponds an participates in the benefits when they are drained and a protei supplement becomes available. This kind of project, replicated in man parts of the world, develops thorough environmental understandin from the base up, and on the sound foundation of understanding amon the local communities.

Environmental understanding is also being stimulated by many non governmental movements. In Uganda, the Association of Wildlife Clubs is active in promoting environmental education in schools, while other

NGOs in Kenya are advising on the environmental content of a new school curriculum. In India the efforts of the Chipko Movement in protecting trees from destruction have achieved worldwide publicity, and there are many local schemes of reforestation in the eroded and degraded hill lands of India and Nepal. In Mali, in the Sahel, an education project is reaching out into the schools, and creating clubs designed to stimulate local discussions among the students themselves about the future of their environment. Such enterprises as these have been repeated many thousandfold in the developing world.

This building from the grassroots up is a most hopeful approach. It leads individuals to recognise that they can do something both for themselves and for the communities of which they are a part and that they should, indeed must, act in this way if their children and grandchildren are to have a satisfactory future. Not surprisingly, this approach has more appeal to its participants than that which has had to be adopted in the societies of the developed world which are more detached from the land. However, the example set by the developed countries to the developing world establishes individual wealth, consumerism, and personal prosperity as goals, although it is the pursuit of these values which threatens the same environmental degradation as has been caused in the developed world. Action at village community level in the developing world must be complemented by action to build individual awareness among the affluent sectors of those societies, and to influence their industries and governments.

There is one issue in the developing world which above all others needs to be tackled through the development of personal awareness. It is the need to limit the present rapid growth of human populations. In many countries these are increasing at between 2 per cent and 4 per cent per annum, and placing an ever-increasing stress on finite environmental resources. The reasons for this growth are well known, ranging from a sheer love of children and pride in a large family, through to the need for a strong family work force and support for the old. The solutions clearly lie in the provision of financial security, health care and family planning facilities to the millions of women who wish to limit their fertility but have no means to do so. But public awareness of the need to constrain population growth must obviously be developed in a sensitive manner, given the intimacy of the personal choices involved. Information that tells people, especially among the rural and urban poor, about the means and benefits of planned parenthood is vital: it should set the matter within a wider frame of social care and social policy so that birth control becomes widely acceptable in the community. This is the aim of many national and international campaigns using the media of televisio documentation and personal advice, including those of the Internation Planned Parenthood Federation and the United Nations Fund f Population Activities.[17]

TOWARDS MORE SENSITIVE PERCEPTION

Much of the world's environment is degraded by the action individuals, and can only be healed by individual action. In many par of the world environmental damage is being caused by activities th individuals could easily stop. Opinion polls in many areas have show concern about noise, water pollution, foul air and litter. Noise is ve largely generated by individual behaviour: the youth who asserts h dominance by riding a noisy motor bike or the owner of a radio blari to the common annoyance. It is perfectly possible for consumers demand quieter motor bikes, chain saws, cars and commercial aircra The quietening of the latter has indeed partly reflected public annoyan at excessive noise levels around airports. The limitations of the use Concorde, a brilliant technical achievement but nonetheless a noisy an polluting one, may be recorded in history as one of the first instanc where a high technology invention had its wings clipped by publ environmental concern. Another such instance is provided by nucle power, where public concern over reactor safety following inciden such as those at Three Mile Island and Chernobyl has led directly to number of countries halting new orders for nuclear reactors and even decisions to phase out this energy source. Concern in Europe about t die-back of forests due in considerable part to the oxidants generated sunlight from motor vehicle emissions is leading, as it has in the Unite States, to demands for the highest standard of emission control, an might even promote individual action to pay for catalytic converte where these are not demanded by law. The problem of litter is, of cours down to the individual: we must all learn to avoid making litter, an deter others from doing so. The key is to make people want to chang their own behaviour rather than devote their energies to campaignin for someone else to stop some anonymous 'them'.

Green consumerism is rising in Europe, the United States an Australia. Books drawing attention to environmentally-friendl products, like the UK *Green Consumer Guide*, have enjoyed substanti sales. An Australian committee has recently published a *Personal Actio Guide for the Earth* which lists actions people can take at home, in th garden, out shopping, when travelling, at work, in the country and within the community to reduce their impact on the environment. These actions need to be encouraged as a kind of personal non-aggression pact with nature. A *Declaration of Fontainebleau* adopted at the 40th anniversary of IUCN[18] places the primary responsibility for the future of the world's environment squarely on the individuals concerned. It states:

> If humanity is to find the right way forward, it must base its advance on a code of values that is less aggressive and more caring for the earth. A code that reflects a deep sensitivity to the ecological interdependence of our planet, and a respect for life in all its forms. A code that draws from all sources of knowledge, all cultures and all memories.
>
> We must, above all, understand that human rights will only be achieved if we respect the rights of our environment. While every human being has the right to a healthy and productive environment, to clean air and drinkable water, that person also has a duty to pass these vital resources on to future generations and to conserve the tremendous variety of life and the fragile balances of the biosphere. On this, the spiritual, cultural and physical survival of our own species depends. The responsibility is individual, collective, and inescapable.
>
> Political, economic and social leaders must work together to ensure that the world's natural resources are shared by all humanity and are no longer the subject of wasteful over-consumption by a minority. Development that destroys, erodes and pollutes must be replaced by a sustainable development which protects the quality of the soils, airs and waters and maintains the diversity and the productivity of the lands and the seas. This is the only way to ensure the well-being, dignity, and fulfilment of both present and future generations.
>
> For caring for nature also means caring for people. We must break out of the vicious circle through which the degradation of the environment and poverty are inextricably linked. We all share a responsibility for relieving this distress and establishing in its place an enduring harmony between human populations and natural resources, drawing on science and education within a social context that is constructive.
>
> To succeed, our conservation strategies must draw on the energies of every nation, from the remotest village to the capital city. They must take full account of the riches of experience and wisdom of all traditions and cultures. They must have at their heart the

disadvantaged inhabitants both of the countryside and the cities. And they must give proper recognition to women, who often bear the heaviest burden in the development process.

We call upon men and women in all countries to sign this Declaration as a demonstration of their determination and solidarity, thereby concluding a non-aggression pact with nature.

It is clear that the future of the world's environment depends on how far collective awareness of our problems is reflected in individual action, and how far this in turn feeds through into collective action by local communities, industries and governments. The basis of our actions must be a shared ethic, a common belief that our responsibility to our children and grandchildren demands that we nurture the world's environmental life-support system. We should also recognise that that life-support system is made up of a great diversity of species of plants and animals, evolved like ourselves in interplay with the physical and chemical features of the world, and who have a right to exist there. There is an ethical reason for concern for species other than our own, as well as the very practical one that we depend on them.

This attitude and this recognition of shared responsibility needs to be developed within the many sectors of today's complex societies. At the individual level, we have the following needs:

1. *Informed media* that give balanced, unexaggerated, non-trivialised information for the public at large, and emphasise the responsibilities and opportunities for individual people to take action that will bring environmental benefit.
2. *Advice*, to help people in particular situations to build a sounder environment, to plan their families, and otherwise achieve the goals they may seek but do not know how to attain.
3. *Education*, at both formal and informal levels, directed at all sectors of the community. In the developing world a great deal can be done by informing the women who themselves educate the next generation and manage many of the environmental resources of the community. We can do much by reinforcing the traditional values of communities living on the land, and strengthening their recognition that these are inherently sound and need adapting to the modern world rather than casting aside.
4. *The arts*: much art has been inspired by nature down the ages and art in many forms can contribute greatly to our care for the environment. Beautiful photographs and wonderful films or

television have done much to move people to environmental sensitivity. The creative arts, like great landscapes, combined with the availability of information centres on the environment, can show something of the diverse inspiration that nature gives to the human experience. The environment is about more than survival and science.
5. *Religions*: most of the world's great religions emphasise reverence for the creation, and a duty of care towards our fellow human beings and indeed towards the wealth of nature with which we share this planet. In today's world there is great need to recapture that insight and vision. The world needs prophets today as much as it ever has done.

Of course action needs to be taken at the group level as well. This is the concern of other chapters in this volume. But group activity mirrors the individual. It demands leadership by caring, informed and persuasive people, supporting the commitment of individuals. It needs support from services ranging through family planning and health care into environmental management, agricultural support, the development of responsible industries, and the provision and advertisement of goods that are friendly toward the environment.

Both individuals and groups need access to new discoveries and insights. We need science in order to understand the environment better. We need to develop economic methodologies that ensure a true evaluation of the resources of the environment when judging the best route for development. We need techniques for assessing the potential impact of industrial and other developments on the environment. We need understanding of how climatic change may alter the sustainability of national plans. All these things need to feed back into personal and public guidance, so as to avoid a waste of human effort and resources and an aggravation of suffering.

Governments have an indispensable role to play: to ensure that all these activities actually take place. They also need to recognise, more than they have so far done, the inter-sectoral and inter-disciplinary demands of the environment. They also need to accept the crucial importance of the international dimension, and the duty of collective care in relationships between states.

Finally, concerned citizen groups are the conscience of many states. They need to act fearlessly to promote what the enlightened individuals in their communities believe should be done. They have a part as prophets, and the world needs prophets even if it disagrees with them.

Starting from the basis of taking responsibility for one's own life and for one's actions towards one's neighbours and the environment, we all need to participate in building structures that will make the collective human system more responsive. Only by the majority of individuals assenting to the need for environmental care and sustainable development will this latter happen, and this will not be easy for everyone. It may infringe certain traditional freedoms and deeply-held beliefs. For example, without population control there will be no environmental future or decent standard of living for the bulk of humanity, and people need to see that they should limit their fertility in a responsible way, even if this leads them to depart from the traditions of their ancestors. Again, unless we can bring a high quality of life to the poorer parts of the world, their sustainable development and the balance between the numbers of their people and the environment that supports them will be jeopardised. This, in turn affects the attitude and quality of life of people in the developed world.

People need reassurance. They can have a high quality of life, productive employment, and a beautiful environment if they are prepared to adapt their lifestyles as individuals, and as collective groups. They need to know how to do this. Too many strategies have been built from the top down. We need to build much more from the ground up, from the individual to the community. We need more investment at local level, and more respect for the wisdom of traditions. But above all, we need recognition that tomorrow's environment is the responsibility of everyone today, and that each person's actions are the bricks of which that edifice will be built.

9 Religion and the Environment

David Gosling

INTRODUCTION

In an earlier chapter, Sir Shridath Ramphal maintained that before sustainable development can become established as a global working principle there will have to be 'a transformation of attitudes in some fundamental respects . . . ', a 'recognition that we all have an obligation to future generations as well as to ourselves . . . ', and a 'need to see environmental problems in interdisciplinary terms . . . '. He also argued 'that we must think of our planet not only as a world of many states but also as the state of our one world . . . '.[1]

Of these four prerequisites, the first, a transformation of attitudes, is legitimate territory for religions. The second clearly falls within the accepted prophetic traditions of Judaism, Christianity and Islam (with the exception of certain fundamentalist groups which preach the destruction of the world as a consequence of human wickedness). The third, the need to study in interdisciplinary terms, has been increasingly recognised by those who are willing to understand religion as a universal phenomenon; and the willingness of major world religions to think globally, the fourth prerequisite, is evidenced by the fact that all of them are involved in various international and ecumenical religious and secular bodies committed to global solutions to common problems. Thus, for example, the United Nations Environment Programme (UNEP) receives periodic input from all major world religions, and the Festival of Faith and the Environment held in Canterbury in September 1989 included major contributions by Christians, Buddhists, Bahais, Muslims, Hindus, Jews and Sikhs.[2]

Martin Holdgate has specifically mentioned the world's great religions as resources for the promotion of

> reverence for the creation, and a duty of care towards our fellow human beings and indeed towards the wealth of nature with which we share this planet. In today's world there is a great need to recapture that insight and vision. The world needs prophets today as much as it ever has done.[3]

In practice religion has often heightened the aggression and selfishness which Martin Holdgate deplores! In the next section some misconceptions about religion will be discussed. This will be followed by details of some important recent religious events relating to the environment which took place in Europe. Finally, two fruitful religious ways forward will be summarised, one from a Buddhist, the other from a Christian perspective.

SOME MISCONCEPTIONS

It is often supposed that in ancient times, when religions tended to be more influential than now, the relationship between human life and the rest of creation was harmonious, this relationship having subsequently been destroyed by secularisation and technological and industrial progress. At some times and in certain places this may have been the case. But the ancient world of Greece and Rome, for example, was responsible for an enormous amount of deforestation around the Mediterranean basin. Even the Jews, with their careful agricultural regulations governing such matters as the jubilee year when land was left fallow in order to recuperate, were less than meticulous in other respects. Thus the majestic cedars of Lebanon, praised for their strength and longevity in the Psalms, were soon cut down to fuel innumerable wars and to provide wood for the restoration which followed them. And although the Jewish Scriptures are often cited (usually by Christians) as reflecting a society which treated the environment in an ecologically sensitive manner, there are no grounds for maintaining that the Jews were any better in this respect than, say, the Indus Valley civilisation which preceded them by a thousand years.

Christians often point to passages in the Old Testament which appear to suggest that a day will come when an exceptional state of peace (*shalom*) will prevail, when wolves will cease to prey on lambs. But there are no clear-cut reasons to suggest that such a truce was ever conceived as more than temporary. Our distant ancestors probably viewed 'nature' much less sentimentally than we do today, at least in the West. But if relationships between human life and nature in the past were much less harmonious than is often supposed, those of more recent times may be much more complex conceptually and in practice, differing from continent to continent. Two examples, from Europe and Asia, will be summarised to illustrate such differences.

When modern Christian theologians try to address themselves to the problems raised by science, technology and the environment, they tend to be faced by two major sources of difficulties. The first is the legacy of Kantian thought, with its limited notion of what constitutes moral behaviour and activity. The second is the Western tendency of making unnecessarily sharp distinctions between God, humanity and the rest of the universe. The former is fairly domestic in that it reflects a particular religious and philosophical tradition in the West, whereas the latter relates to a much broader range of Greek and Judaeo-Christian thought.

Kant believed nature to be a collection of irrational forces which needed to be subdued and kept in check by human effort. 'Man' was a rational and spiritual being whose holiness was associated with his moral personality, and part of his moral duty was to subdue nature. Thus the world of morality, with its inherent possibility of holiness, was sharply distinguished from the world of nature.[4]

The ideas of Kant and of many Christian theologians influenced by him illustrate two essentially Western theological presuppositions which seem to prevent modern theology from finding a way out of the difficulties described above. First, the creator God is absolutely world-transcendent, nature is absolutely non-divine and the only place for divinity is God in his non-worldliness. Second, there is a sharp qualitative distinction between humanity and non-human nature which gives the former 'spiritual' freedom and reduces the latter to an absolutely subordinate role. Thus nature is the arena for our spiritual freedom and its worth is purely instrumental; no moral limits are imposed on humanity with regard to the use of nature; and there are assumed to be no intrinsic internal limits within nature that we should respect.

That nature does, in fact, possess internal limits of its own is becoming increasingly apparent after decades of abuse, the consequences of which are described in other chapters in this volume. When history books are written it is likely that the present century will be stigmatised as one in which priceless and irreplaceable chemical and mineral deposits and genetic species which took thousands of years to form were squandered by a handful of naively expansionist and consumer-orientated governments. It will have to be sadly acknowledged that their mistaken view of progress was often underscored by interpretations of Christianity which were woefully misguided.

According to most Asian religious traditions, God (or whatever entity stands in the place of God), humanity and nature flow into one another rather along the lines suggested by Martin Holdgate. But in India, for example, this did not make much difference to the manner in which an essentially Western pattern of progress was experienced, and Prime Minister Nehru's commitment to rapid industrialisation fostered dependencies which in the long run have been weakening both to human communities and to their habitats.

Whereas Indian society as a whole suffered from patterns of 'development' which achieved little of permanent value, the encounter between India and the West also fostered reformulations of religion which provided illuminating insights into the problems of social and environmental degradation and ways of solving them. Gandhi, for example, who at the end of Nehru's premiership was considered to have been little more than an idealistic dreamer, has come to be recognised by many as a prophet whose vision of village 'republics', based on the essentially Hindu doctrine of *sarvodaya* ('the awakening of all'), might have done much to break the neo-colonial patterns of dependency into which many people found themselves locked, as loans, escalating fuel prices, and tariffs took their toll. Gandhi had preached an unusual blend of ancient and modern wisdom – home-spun cotton, asceticism, *satya-graha* (truth insistence), *ahimsā* (non-violence – essentially derived from Jainism), decentralisation and participatory government (except where he himself was concerned!). In retrospect, might it not all have added up to a viable alternative to the grand white elephants of the new Raj and its Western acolytes? 'What do you think of Western civilisation?' an admirer once asked the Mahatma. 'I think it's a good idea,' came the tart reply.

The way forward to a unity of perspective incorporating what we call the religious and the secular into a new framework seems to be a formula whereby religion offers much to the improvement of our world in every aspect including the state of the environment, and the work of M. K. Gandhi is a potent example of such an approach. Gandhi's efforts were not directed primarily at environmental problems, but his integrated vision implicitly involved a more benign attitude to nature than that of governments and planners. It was not enough that India possessed a religious tradition in which all life and nature were revered. It needed a Mahatma to set the philosophy within the dynamic of lived ideas and images which captured the imagination of the public at large.

RELIGIOUS PRESUPPOSITIONS

A Harvest Festival in Cambridgeshire in October 1988 featured the following: an agricultural exhibition which highlighted ways in which modern agricultural methods – including the intensive use of chemical fertilisers and pesticides – are used to maintain an aggressively affluent lifestyle; prayers thanking God for his goodness as represented by enormous crops of just about everything; performances by the Royal Lancers and Royal Fusiliers; and a hymn which combined the praise of military success, earth's bounty and the well-being of future generations, attributing all three to divine providence:

> Thy strength made strong our fathers' hands,
> A people great on seas and lands,
> To win, till earth shall pass away,
> Such honour as the earth can pay.
>
> O God, whose mighty works of old
> Our fathers to their sons have told,
> Be thou our strength from age to age,
> Our children's children's heritage.[5]

The anthropocentrism of these and other verses in the hymn is essentially that of a small child who relates the world to himself and his country and expects God to intervene periodically to provide victory in war and bountiful harvests. The underlying idea of a God who made nature rather like a huge clock at some time in the distant past, and then left it except for occasional interventions or 'miracles', is known as deism, and enjoys considerable popularity, especially among fundamentalists and others who explain the 'miracles' as signs of divine favour or disfavour according to preselected scriptural texts and/or moral prescriptions. Such an approach induces feelings of inferiority and guilt, and presents a picture of God which is diametrically opposed to the loving Father of the New Testament, whose rain falls equally on the just and unjust, or the compassionate and merciful God of Islam.

In spite of the unfortunate connotations of the Harvest Festival presented above it is important to recognise that mainstream historic Christianity, of which Roman Catholicism, Orthodoxy, and Anglicanism are examples, has refuted deism in favour of more integrated and continuous relationships between God and the Universe (panentheism).[6] The Western Christian discontinuities between God, humanity and nature are nonetheless sharper than those of the Indian religious tradition or in the comparable organic models of feminist theologians such as Sally McFague.[7]

The view that human beings are in some sense co-creators with God and that creation is an open process for which we share responsibility for the future is a theological position which brings theistic religions closer to, say, Theravada Buddhism, which is, strictly speaking, non-theistic, than to the ideas of fundamentalists who believe in a manipulative God who is quite separate from both humanity and nature. It is therefore not surprising that when Christian environmentalists and representatives of other living faiths join forces, as happened in September at the Festival of Faith and the Environment at Canterbury, the fundamentalist wing of the British churches is in strong opposition. For the reasons just given, the divergence between mainstream Christian churches and denominations, with their potentially inclusivist approach to the whole of creation, including people of other living faiths, is much smaller than the gap that separates them from exclusivist fundamentalism, with its primitive and essentially tribal notion of an interventionist God.

NEW BEGINNINGS

In September 1986 the World Wildlife Fund marked its twenty-fifth anniversary with an interfaith ceremony at Assisi. The five religions represented were Christianity (the Minister General of the Franciscans attended); Tibetan Buddhism; Hinduism (via the President of the Indian Virat Hindu Samaj); Islam (the Secretary-General of the Muslim World League); and Judaism (the Vice-President of the World Jewish Congress was present). There were five separate liturgies, each of which contained a statement about the relationship between religious faith and the Earth and the importance of living faithfully. The event represented the inauguration of a new network on religion and conservation and marked the beginning of a process of building an interfaith community with a common cause.[8]

The recent Canterbury event combined a conference on Christian Faith and Ecology jointly sponsored by the World Wildlife Fund and the British Council of Churches with a simultaneous Festival of Faith and the Environment. Seven major world religions were represented, the Bahais having joined in 1987, and the Sikhs at the Canterbury event. Pilgrims from these seven religions were welcomed into Canterbury Cathedral at the beginning with a short service of psalms, prayers, a reading from Proverbs and an anthem, and paschal candles were symbolically lit to represent the spreading of light among all present.

The gathering as a whole was marked by humility, cooperation and the need to grow through differences via mutual sharing and concern for our world. According to Ruth Page, 'God feels joy when diversity is working'. There was an acknowledgement that blindness must be recognised before a way forward can be discovered. The various religious acts of worship followed consecutively, each 'for the faithful of that faith'. But all were welcome to share in one another's celebration.[9] Sunday worship took the form of a Creation Eucharist, which included some interesting material such as the recasting of the traditional Benedicite as a lament for our despoiled world:

> And I, the earth the Lord created,
> Cry to him who made me,
> Save me from ravage and destruction
> To praise and glorify thy name forever.[10]

In his address the Archbishop of Canterbury told the congregation:

> The conviction that nature does not exist simply and solely for the benefit of humankind . . . is becoming increasingly widespread and articulate. Because it finds its true source at such deep levels of the human spirit, it must, I think, be called a religious conviction. But it is not a conviction unique to any one religion in particular, and it is shared by some who would profess no religion at all.

Strong opposition to the Festival and to the hospitality afforded to it by Canterbury Cathedral was expressed by fundamentalist groups, and a nationwide 'March for Jesus' appears to have been staged deliberately to coincide.

The Assisi and Canterbury events were major occasions at which religions affirmed the importance of the environment and urged their participants to new and renewed commitment. There was also an imaginative Creation Harvest Liturgy in Winchester Cathedral in October 1987 at which representatives of Buddhism, Bahai, Hinduism, Islam, Judaism, Sikhism and Taoism were present.[11] On 5 June 1986 (World Environment Day) the World Council of Churches hosted a United Nations Environment Programme event at which Muslim, Jewish and Christian world religious leaders pledged themselves to 'preserve the integrity of creation', and a tree was subsequently jointly planted at the UN.

UNEP produced its own liturgical materials for World Environment Day the following year.[12] The Catholic Fund for Overseas Development (CAFOD) has produced some excellent liturgical and study material which emphasises the need to see environmental and development problems as justice issues.[13] The United Reformed Church has very recently produced a beautifully-illustrated anthology of prose and poetry, prayers and hymns on the theme of creation and the environment, and the Methodists have followed up their widely-circulated tape-slide 'Making Peace with the Planet' with worshi material designed to explore the urban and seafaring environments o Liverpool and the Shetlands.[14] The British Council of Churches an Christian Aid have put together a resource pack based on the churches 1983 Vancouver call to covenant for Justice, Peace and the Integrity o Creation.[15] The Tear Fund has produced materials for children an some imaginative display materials suitable for harvest exhibitions. But some of the most challenging materials, often unavailable i English, are from parts of the world where the environmental crisis bite most acutely. For example, a Filipino version of the Stations of th Cross for the Death of a Forest attempts to describe the tragi consequences of ecological devastation upon the physical and spiritu lives of the poor. The irony is graphic:

> Jesus turned to those weeping and said: 'Weep not for me but f yourselves and your children, because the day will come when t people will say to the mountains, "Fall on us", for if such things these are done when the forest is green what will happen when it is desert?' . . . After the burial of Jesus there is a fifteenth station: t Resurrection. After the death of the rain forest there is resurrection. The forest will not return to life.[17]

These are some recent attempts by churches and denominations express the view that creation is one and that religions must share th resources in order to explore appropriate responses to the current cri Collectively they represent a move away from anthropocentrism t belief that we must somehow grow into God, and that, in spite of exploratory nature of the quest, we must act decisively now. There i move from individualism to interdependence, and nature is increasin being recognised not as a neglected appendage to humanity but as integral component of the seamless fabric of which we are a signific but not unduly dominant part.

ENLARGING THE FRAMEWORK

Two religious expressions of environmental concern, which are in line with the trends of the previous section and the aspirations of Sir Shridath Ramphal and Martin Holdgate quoted at the beginning of this article, come to mind. One is taken from Buddhism, the other from Christianity; both are fairly tentative and exploratory.

For the last two decades Buddhist monks in Thailand have become increasingly involved in a variety of developmental activities which have gone hand-in-hand with a reinterpretation of Buddhism. Most Thai monks come from poor provincial rural backgrounds and ordain at an early age, subsequently using the monastic educational system to migrate via provincial capitals to the metropolis, where the most able obtain degrees at Mahamakut and Mahachulalongkorn Buddhist Universities. As part of their degree courses they are sent back into the provinces to assist villagers with community enterprises such as the construction of rice and buffalo banks, irrigation, handicraft production and many other activities in which women now play a major role. The Thai Government initially encouraged and financed these programmes because it thought that they would help to counter insurgency along the Kampuchean and Laotian borders and provide alternatives to opium production in the extreme north. But the monks became deeply immersed in the social, economic and increasingly environmental dimensions of rural development, and many continued with it after graduation, returning in some cases to their own home areas.

In parallel with this practical concern for the building of self-reliant rural communities has been a reinterpretation of Buddhism to emphasise its this-worldly teachings as expounded by a number of brilliant scholar-monks of whom Buddhadasa (Putatat) is probably the best known. Putatat lives in a forest ashram in the south of Thailand; the essence of his teaching is that the 'no-self' (*anattā*) doctrine of Buddhism relates primarily not to the denial of the Hindu soul so much as to the process whereby, in this life, we remove the individual self which is the root cause of attachment and craving. This we do via consciousness characterised by 'emptiness' (*sūnyatā*), the 'void', or whatever we call the unconditioned which is Buddhism's closest approximate to God as conceived by theistic religions. Such a view comes close to Mahayana teaching with its stress on the figure of the *bohdisattva*, who finally refuses to accept his individual right to enter *nibbanā* until all sentient beings are able to do the same.[18]

From this thumbnail sketch it should be clear that the monks' involvement in development work, which has increasingly meant improving the environmental context of villages, has occurred together with a major restructuring of classical Buddhism in a manner not unlike some of the shifts noted in the last section – the move away from individualism towards community, for example.

The second example relates to an initiative by churches at the 1983 Vancouver Assembly organised by the World Council of Churches to covenant for Justice, Peace and the Integrity of Creation. Subsequent exploration has demonstrated that ecological/environmental issues subsumed under the third section of the theme are at least as important as the other two more traditional ecumenical concerns. Australian zoologist Charles Birch describes the three collectively as 'the momentous instabilities of our time', which are closely interrelated and must be addressed together if our world is to have a future.

This kind of global approach may be unfamiliar to most British Christians, but it invites church members and others in different parts of the world to search for covenants which will bind people together in common causes. The twinning of the cities of Coventry and Dresden after the war was a powerful example of a covenant in the 'peace' area. Agreements between US churches and Nicaraguans following the Vancouver Assembly were a good example in the 'justice' area, but as yet little thought has been given to appropriate environmental covenants. One nettle, however, that is increasingly being grasped by international and national bodies, including the Roman Catholic Church and Christian Aid, is the whole question of debts to the International Monetary Fund and the World Bank, debts based on economic assumptions which fail to take into account the enormous cost of environmental pollution during the past three decades.[19] Such rethinking of the true cost of 'progress' to large sectors of the human population and the environment reflects Shridath Ramphal's call to 'see environmental problems in interdisciplinary terms'; it also challenges many naive assumptions about borrowing and lending which are sanctioned by religion. The Justice, Peace and Integrity of Creation approach also complements Ramphal's plea to 'think of our planet not only as a world of many states but also as the state of our one world'.[20]

These two examples, one from Thai Buddhism, the other from an international federation of churches, are illustrative of the kind of revolution in religious thinking that must come if religions are collectively to be able to address the challenges mentioned by other contributors to this volume. In the prophetic words of the late Mary Clare, former Mother Superior of the Sisters of the Love of God,

We must try to understand the meaning of the age in which we are called to bear witness. We must accept the fact that this is an age in which the cloth is being unwoven. It is therefore no good trying to patch. We must, rather, set up the loom on which coming generations may weave new cloth according to the pattern God provides.[21]

10 Industry and the Environment: A Question of Balance

Christopher Hampson

It has been all too seldom that industrialists have had the opportunity t contribute to the important debate on striking the balance alluded to i my title. The responsibility for this failure of communication does n necessarily rest with the environmentalists. I think my industri colleagues have been somewhat reluctant to speak out and appear c public platforms.

Nevertheless, this failure has also come about as a result of t general public's perception of industry – and of the chemical indust in particular – as the main culprit in causing environmental problem The 1987 poll done by the Chemical Industry Association showed th over two-thirds of the population thought of the chemical indust as a polluter, and over a half expressed concern about the safety chemical plants, figures that have not improved over recent yea despite very considerable efforts by the industry to correct this (our view at least) false impression. It may seem that having representative of the chemical industry address an environmental deba is a bit like asking Genghis Khan to expound the virtues of peace co-existence.

I have taken as my topic, 'Industry and the Environment: A Questi of Balance', because that is, of course, the question with which we mu grapple. No informed participant in the debate is suggesting that should stop all industry to protect the environment, (althou occasionally there are those who say petulantly, 'Of course if you wan perfect environment, you are going to close down all industry'). That not going to happen and we all know it. Similarly, all thinking peo know that it is not possible to have industrial development and have impact on the environment.

Specifically, I was asked to comment on the Brundtland Report on *Our Common Future*. I can put any fears quickly to rest by saying that industry welcomes the recommendations of that Report, which suggests some guidelines for the future and talks about the need to consider environmental protection in the light of 'sustainable development'. That is easy to say; the hard part is really to decide how this should be done, and what the role and responsibility of industrial companies such as mine is in helping strike the 'right' balance between industrial development and environmental impact.

In looking at these problems, I would like to make three points. First, industry is not insensitive to the need to improve its environmental performance, and the problems are being tackled in a more responsible way than the public might judge from what it reads or hears. Second, a great many of the threats to the environment which have been perceived can be dealt with, but this can only be done if society is prepared to pay the cost. To suggest these improvements can be achieved at no cost to society (in lifestyle, community progress, or otherwise) is unrealistic and does not advance our common cause. Third, there is perhaps more of a common agenda between industrialists and environmentalists than both parties realise. There is, however, some danger that if this common agenda is not recognised and acted upon, excessive bureaucratic intervention may stifle progress. The common agenda approach could best be achieved through constructive debate rather than both sides operating in the confrontational mode. Let me turn to the first of these points.

There is no doubt that mankind's unique abilities over all other animal species – our powers of reasoning, our manual dexterity and our inventiveness – have enabled us to influence our environment, and to shape and mould it towards achieving our own ends. Ever since the discovery and utilisation of this planet's natural resources, these abilities have enabled mankind to generate wealth and improve our material existence steadily over that of our forebears. But this progress has not been without cost. Economies built on the exploitation of resources become reliant on further exploitation and development. Once one society has made progress, other societies naturally want to emulate that success and undertake their own development. In the past, this continuum was carried out with little or no concern for the long-term impact on the environment.

But the rate of economic development and the consequent impact on the environment has accelerated rapidly over the last 100 to 150 years. In this more recent time span, we have seen the impact of the availability of greater amounts of energy, the development of industrial machinery and the impact of the industrial revolution, the proliferation of industrial and private transport, heavy industry and light industry. The world has moved on from the massive exploitation of natural resources to the extensive use of man-made goods based on natural feedstocks of coal, and later oil and natural gas. Many of these advances were a direct result of developments in chemical process.

In retrospect, the consequences are all too visible. We can all now see that industrialisation inevitably has an impact on the environment. However, in those early days, environmental impact studies had yet to be invented and industry was mainly concerned with the exciting new development potential and immediate wealth creation. Until that wealth was created, society did not have the luxury of worrying about the environment. The plants and processes were themselves inefficient and wasteful. They generated high levels of waste, so that pollution was severe. Waterways and river estuaries became heavily contaminated, not just from industry, but from the tide of urbanisation which flowed in industry's wake. Air pollution became widespread. Smoking chimneys were a sign of wealth and in fact, we still use the expression for home development of 'smoke up our own chimneys'.

We have of course, recently 'discovered' the greenhouse effect, but awareness of the possibility of climate change is not new. In 1896, a Swedish climatologist suggested that the average temperature of the world would rise about 4 degrees centigrade within a few centuries. In fact, it appears that the earth's average temperature has risen about 1 degree centigrade since the 1800s, some would say because of our large production of CO_2 through burning fossil fuels to feed the fires of industrialisation and modern every-day living.

However, the evidence of cause and effect is far from clear. By examining cores of Antarctic ice that date back 160,000 years, scientists have found that ice ages have coincided with reduced CO_2 levels, while interglacial warmings have been marked by an increased production of the gas. Thus, industrialisation is not the sole factor at work. Roughly half the world's emissions of CO_2 are absorbed by sea and land. Forests play a vital role in absorbing CO_2, but in areas such as India and Brazil, millions of hectares are still being destroyed annually to provide agricultural land. In fact, smoke from land-clearing fires accounts for about 20 per cent of the world's 5.5 billion tonne CO_2 production.

Similarly, recent increases in atmospheric levels of other greenhouse gases cannot be directly attributed to industry. Take methane, for example, levels of which have increased by 1 per cent per annum over the past decade. Major sources of this gas are municipal landfill sites, rice paddies and even flatulent cows! Indeed annual emissions of methane from these last two sources have recently been estimated to be 120 million tonnes and 76 million tonnes respectively.

None of these factors mean we can deny the impact of industrial operations on the environment, and it is a fact that industry continues to suffer in the eyes of the public from its early performance and from the visible signs of its presence in more recent times. Industry does have a key role to play in improving the environment, but it is one which it cannot achieve on its own.

It is in this sense that industry welcomes the World Commission's Report, because the Report does stress the need for society as a whole to decide how the sometimes conflicting claims of immediate material well-being and long-term sustainable growth can be settled. The report correctly recognises that no one factor of society can address the matter on its own, but that there are a whole series of 'interlocking crises' which need to be carefully considered and managed. It also recognises that the continuing role of industry in creating wealth and economic development is essential if the world's most pressing environmental issues are to be tackled with any hope of lasting success. So what measures can industry take in its role as a component in sustainable development?

While by no means suggesting that all is rosy, and that you can therefore relax in the certain knowledge that industry will solve its own environmental problems, I would like to suggest that large segments of industry are in fact reacting much more sensibly and sensitively than you might think on the evidence presented by the media. Any manager in an industrial operation today is well aware that the environmental impact of his business, as seen by the community, is a factor of paramount importance in his ability to stay in business. Industry has a 'licence to operate' granted by the community either explicitly, or, more often, implicitly. An industrial facility which makes the community feel that the benefits of the continued operation of that facility, in terms of wealth creation, jobs, community considerations, is outweighed by the harm caused to the environment, loses its licence to operate and gets shut down. That can apply as much to an industry as to an individual plant. We all understand that, and there is a tremendous effort going on to clean up our performance, which manifests itself in a number of ways: considerable effort, investment and training to improve actual environmental performance; new technologies and processes to improve efficiency and reduce environmental impact, for example, by not producing wastes in the first place; and greater efforts on community relations and understanding community concerns.

Forgive me if I quote at least one example from our own experience in ICI. Throughout this century, the River Tees in England has been a major industrial artery and the economic life of the region still depends to some extent on it remaining as such. The growth of the iron and steel industry in the early part of the century, followed by the development of the chemical industry, resulted in the fish life being virtually eliminated from the river by the mid-1930s. Continued industrial expansion, particularly by the chemical industry, and the discharge of untreated sewage resulted by the late 1960s in the Stockton/Middlesborough stretch of the Tees being completely devoid of oxygen for long periods during the summer months. It was around this time that ICI, along with other industry and Northumbrian Water, realised that steps had to be taken to reverse this trend. It was recognised that any improvements could only be achieved gradually, otherwise whole sections of industry could face total shutdown with disastrous effects on the economy and the community.

A joint programme was therefore set up which has markedly improved the quality of the river, while still taking into account the needs of industry and the country. During the 1970s, for instance, there was a steady improvement in condition of the river and ICI halved its discharges. Since 1970, ICI discharges into the Tees have been reduced four-fold from 480 tonnes a day to about 120 tonnes a day. This has been achieved by ICI spending millions of pounds replacing old plant designs with more efficient new ones, the installation of abatement equipment and improved operation control. There has been a dramatic reduction of liquid effluent, with similar dramatic reduction in other major pollutants such as ammonia, mercury and cyanide. The results of this long-term effort are now paying environmental dividends. The number of reported catches of salmon in the Tees is on the increase, and judging from the regular detailed surveys carried out by ICI's internationally renowned Brixham Environmental Laboratory, and Northumbrian Water, there has been a steady increase in a wide variety of life forms in the river and its estuary. The number of species of invertebrate animals recorded in the river Tees more than doubled between 1978 and 1985.

Unfortunately, as this example demonstrates, the conception, development and introduction of new technology takes time and money. Public awareness of environmental matters has been growing over several years, but more recently has accelerated rapidly. Government awareness has undergone a similarly rapid change. Industry has found it difficult, if not impossible, to respond at the same rate. It is not purely a matter of capital investment and economics. The world has come to depend in important ways on some products or processes which are not environmentally perfect, and this factor must be taken into account as well.

Another example of the importance attached by industry to environmental matters is the articulation by a great many companies of an environmental policy, promulgated with the firm backing of senior management. An example is that of 3M, as reported in an article in *Management Today*:

> 3M will continue to recognize and exercise its responsibility to:
> - solve its own environmental pollution and conservation problems.
> - prevent pollution at source wherever and whenever possible.
> - develop products that will have a minimum effect on the environment.
> - conserve natural resources through the use of reclamation and other appropriate methods.
> - assure that its facilities and products meet and sustain the regulations of all local environmental agencies.
> - assist, wherever possible, governmental agencies and other official organisations engaged in environmental activities.

The application of this policy has enabled 3M to reduce waste significantly and improve energy efficiency. The policy statement is in fact remarkably similar to that of my own company, and very probably a great many others as well.

I hope, therefore, that I have been able to show that industry is taking its responsibilities seriously, recognising that its future lies in improving its environmental performance. Many companies have understood that good environmental practice can not only help make them successful, high-quality, efficient manufacturers, but also is an investment in the future. We are attempting to build good environmental practice into all our operating standards and our planning, and this is having considerable success.

Like everyone else, we would occasionally hope to have this good performance recognised. Human nature is such that people often respond better to a pat on the back and encouragement to do better than to a kick in the backside and being told that the job done is unsatisfactory. Perhaps the same is true of industry. I am not disputing that occasionally the debate must be sharpened by confrontation, but I believe we could make better progress if both environmentalists and industrialists could occasionally display some understanding of each other. Exaggeration of positions to make a point can be counter-productive, and destroys public credibility.

The second issue I would like to address is that of cost. So far, most governments have adopted the line that 'the polluter pays'. I suspect that a lot of people believe that this means that in some way industry is going to bear the costs of environmental clean-up without increasing the price of its products. In fact, of course, any cost borne by industry eventually gets passed on to the public in the form of higher prices, or leads to reductions in other valuable enterprises, for example, investment or development. I believe that most industrial environmental problems are soluble – the problem is the cost.

In this context, it is interesting to note the outcome of a Gallup Poll carried out on behalf of the *Daily Telegraph*, concerning environmental matters. Respondents were asked to reply to the following question: 'Sometimes measures that are designed to protect the environment cause industries to spend more money and therefore increase prices. Which do you think is more important, protecting the environment or keeping prices down?' The results are very interesting:

TABLE 10.1

	1982	*1985*	*1988*
	%	%	%
Protect the environment	50	60	74
Keep prices down	40	23	17
Don't know	10	17	9

On the face of it, this is encouraging. However, there is not a large body of evidence as yet to show that the general public will actually pay more for products from companies which have good environmental performance than from those who do not. However, there is a growing 'Green Consumer' movement, whose followers will apparently pay extra for environmentally pure products. So the messages are filtering through. Again, I believe it is important to stress that so far, it is only wealthy societies that can afford the luxury of a 'Green Consumer' movement. If one's concerns are more basic, and life is more a struggle to obtain the necessities for survival, environmental purity may not be high on the list of desirable product characteristics.

The issue of the bearing of costs is highlighted by the history of CFCs. Just about everyone has now heard of CFCs and their possible effect on the ozone layer, or on the greenhouse effect. The theories on the effect of CFCs in the ozone layer were first put forward in the early 1970s. The effect was not proven and in fact, early attempts to check the theory with scientific evidence were not successful. Nevertheless, ICI and other producers of CFCs started work on finding alternative products that would not have the same potential environmental effects.

Such new products were identified and it was clear that they would be more expensive to produce than the CFCs in use. There was frankly no customer interest at all in converting to these more expensive alternatives and so the company ceased its development programme. More recent scientific work has reinforced concerns about CFCs and the atmosphere, and resulted in the Montreal Protocol and international government and public pressure to find alternatives. I might add that we supported the scientific investigations financially and scientifically, and are much in favour of the Montreal Protocol and the further review work that will now take place. There is now a demand for environmentally friendly refrigerants and blowing agents. Our research and development programme has now been reinstated and is running in high gear, and we hope to have new products available in the next few years. They will cost considerably more than the existing products, but the public is now willing to pay that price.

Another example is the use of agrochemicals. Today, we hear of increased concerns about the use of pesticides and fertilisers, to the extent of calls for a complete return to organic farming. It is important to examine the facts. Food cannot be produced in sufficient quantities to meet the needs of everyone without the use of agrochemicals and, in general, so-called organically-grown food will be more expensive to produce. Only a limited number of consumers are prepared to pay extra for organically-grown produce. Moreover, it is possible to demonstrate that organic farming can cause more nitrogen run-off than the proper use of synthetic fertilisers. And of course, consumers have shown no inclination to buy blemished fruit. More importantly, without the same chemicals, the chances of the world being able to produce sufficient food to feed the exploding population are very slim indeed. In fact, in January 1989, the Food and Agricultural Organisation of the UN reported that because of drought there will be a need for a 13 per cent increase in the production of cereal this year to bring world supplies above danger level.

What is needed surely is the proper use of fertilisers, and the development of new agricultural chemicals which can be both effective in their use and environmentally safe. In fact, in the agrochemicals sector, intensive effort and a great deal of money is spent on developing these products.

It may take up to ten years before a laboratory compound becomes a marketable product. During this time, there are not only the hurdles of synthesising the analogue compounds, stringent glasshouse and field tests, and of toxicology but also of process development, plant design and formulation work. There is another whole body of work which traces exactly what happens to the compound once it has served its agricultural purpose. How does it break down? What are its metabolites? Can it be picked up by birds and animals? What residues are left? and so on. To enable industry to develop and introduce a new product, the cost of all this work, which could reach £50 million, has to be reflected in the eventual price of the product. Industry is not only responsive to environmental requirements; it has the ability, given time, both to solve environmental problems and produce environmentally-safe products. The key factors are the community's willingness to pay the price and recognition that such changes cannot take place overnight.

Environmentalists, governments and the public must be understanding of industry's efforts, just as industry needs to recognise the legitimate concerns and aspirations of these other bodies. Changes towards new technologies and new products which have improved environmental qualities take time, and will need changes in the attitudes of society. It is often the people who only yesterday were demanding products then seen as essential to a normal lifestyle, who now demand an instant replacement because of a perceived threat to man's existence. All too often, industry finds itself caught in the crossfire of argument. And not just on products for our home market. The interests of developing nations are sometimes different from the affluent world. The question of balance – benefit versus cost – can sometimes be different, based on a nation's particular economic perspective. Costs to society accepted in one place must mean that something else must be sacrificed.

As I have said earlier, in some ways, environmentalists and industrialists have a more common agenda than perhaps either party cares to recognize. Certainly, only when the public is sensitised to the need for change and prepared to pay for it, is industry's task of bringing it about possible.

And that brings me to my third point, the need for a common agenda and the threat posed by excessive bureaucratic intervention. I have already alluded to one area in which industry and the environmentalists have a common agenda. There are others. For example, we need to ensure that we are working on the right problems and that the public is sensitised to the right issues. If not, it is possible that effort will be wasted. Sometimes, this can result in governments and regulatory bodies also following less productive tasks than they might.

The backbone of corporate environmental policy is the law of the land. Industry makes every effort in its environmental policies to ensure that facilities and products meet and sustain the regulations of all local environmental agencies and initiates guidelines for product testing, residue levels and environmental studies. It is worth noting that the type and amount of pesticides used by farmers and growers is changing dramatically as a result of product substitution and as a result of more stringent regulatory standards. In the last five years there has been a one-third reduction in the tonnage of pesticides being used. International and national standards governing the maximum residue levels of pesticides in food are well based, having been derived from industrial and scientific research.

The same cannot be said about the EEC's Directive governing the maximum admissible concentration of pesticides in drinking water. The reduction to an arbitrary 1 part per billion (ppb) – an amount equivalent to one second in thirty-one years – has not been set on scientific grounds. It simply reflects the advances that have been made in analytical techniques which can now detect chemicals down to parts per trillion. Previously, these very low levels of residues could not be detected and therefore apparently did not exist. Implementation of such limits will impose major burdens on industry and increase the cost of research and development to a point where development could be completely stifled. (It is perhaps worth noting that the British Environment Minister Michael Howard has said that there is no risk to health arising from the levels that have been measured, and that the UK is seeking to persuade the EEC to amend these extraordinarily low quasi-zero levels upwards.)

Another example of stifling bureaucracy is the proposed introduction of the 'VII Amendment' to the EEC's Directive on Packaging and Labelling of Dangerous Substances, which adds an additional and quite unnecessary level of pre-testing and development of experimental products. If this amendment appears in the Statute Book, industry will be charged to carry out costly toxicological studies on even gram quantities of new chemicals before they can be transported from laboratory to laboratory and used in work programmes in the EEC Member States. Unless there is an obvious economic benefit to develop a specific compound at the outset, the innovative capacity of European industry will be severely restricted. The net effect will be to drive research further overseas. As commercial development follows research, this will have an impact on economic developments in Europe in the future. There must be a balance between absolute security and the absolute lack of innovation.

In the Gallup Poll I referred to earlier, 88 per cent of people thought that the government should pass laws to control industry and other producers of pollution while only 5 per cent felt that industry could be trusted to regulate itself. In such a case as the 'VII Amendment', if environmentalists would also say that this was over-regulation, there would perhaps be some chance of success. Similarly, industry at times could improve its credibility by accepting that environmentalists are sometimes right.

As an industry we accept that we should be seen to be well regulated. A proper framework of law and regulations enables all companies to play to the same rules and promotes fair competition. Moves towards the harmonisation of regulations between nations and continents based on meaningful scientific information to produce world-recognised standards are to be applauded, but regulation for regulation's sake is definitely not constructive. As in the issues I have mentioned before, it is a question of striking the right balance between regulation which encourages responsible behaviour, and provides a level playing-field for all, and over-regulation which encourages avoidance of responsibility, stifles development and allocates resources in unproductive directions.

These then are clearly major issues on which I believe we should all welcome debate. For it is only by the frank exchange of ideas and opinions that we will arrive at the right consensus – the 'right' balance.

I remain basically optimistic. I believe most of the issues associated with pollution arising from industrial operations can be solved over time. There are some additional major issues to which I feel we are not, as a society, giving enough attention. With special reference to the concept of sustainable development, let me mention just one, that of energy. The future of fossil fuels as a source of energy is finite (even leaving aside the question of their involvement in the greenhouse effect). Yet replacement by nuclear energy, and the present stage of nuclear technology, is not seen by many as a viable alternative. There is talk of solar power, wind power or tidal power – but these are still concepts and have their own problems. Should not more attention and resources be focused on alternative sustainable and environmentally superior sources of power? Is not this an issue we should all be addressing?

As I mentioned earlier, the Report of the World Commission makes it plain that the move towards sustainable development can only take place through the combined action of all. It is no good if industry is used as a scapegoat for all the world's environmental problems. Industry accepts that it has a key role to play but it can only take part if others will act constructively with industry rather than in a hostile, restricting manner. It is, as I hope to have demonstrated, all a matter of balance and understanding. *Our Common Future* represents a landmark in shaping the world for the decades ahead. Industry welcomes the balanced views presented. If those same views can be implemented through communication and action then the goals of sustainable development can surely be reached.

Even though the path towards more sustainable forms of development will be slow, uncertain and difficult, industry does have a great deal to contribute. It is perhaps unlikely that we will find the perfect solutions to all our problems, or certainly not quickly, but we should not ignore those less-than-complete solutions which will nevertheless have an important contribution to make in the overall plan.

Part Four: National Governments and Sustainable Development

11 OECD Nations and Sustainable Development

Hon. Charles Caccia, MP[1]

> Increasingly the concept of sustainable development is being recognised as a necessary intellectual framework and therefore as a key to tackling major environmental problems.
>
> Sir Anthony Williams, Head, U.K. Delegation,
> CSCE Environment Meeting, Sofia, November 1989.

The Report of the World Commission on Environment and Development (WCED) – a policy focus for the solution of global environmental issues – has generated considerable public and political attention. The fact that in recent times poll after poll shows the environment to be a high priority in the public mind can in part be attributed to the momentum triggered by *Our Common Future.*[2] Politicians are turning 'green' in increasing numbers. They sense a political opportunity. Some governments have taken concrete actions to protect the environment, but few have taken steps to implement sustainable development as prescribed in *Our Common Future*.

Individually, Organisation for Economic Cooperation and Development (OECD)[3] nations are beginning to come to grips with the implications of the Report, but it must be recognised that the ideas contained in it pose many challenges to national governments since they touch upon most aspects of governance: established economic thought; political priorities and horizons; relationships with international agencies, consumers, interest groups (e.g. trade unions, industry, chambers of commerce), and with bureaucracies. The Report also seriously challenges the prevailing and deeply-entrenched attitude which separates economic from environmental goals, since the application of sustainable development will require the integration of economic and environmental policies in budgets, progress and operations. Recent political declarations leave a wide gap between rhetoric and action. A major dose of political courage is now needed to close that gap, given the difficult decisions that remain to be taken. These range from energy-pricing policies to agricultural subsidies and programmes; from land use and urban planning to transport policies; from natural resource exploitation to food production. The obstacles are formidable, at all levels of government and in all sectors of society.

The factors which determine how OECD countries respond to issues raised by the WCED are many, elusive and far from simple. The OECD has launched a number of interesting studies, but so far member nations have not yet arrived at a common set of criteria or principles against which to measure progress. The seven leading OECD nations, known as the G-7 nations, endorsed the concept of sustainable development outlined in the WCED's report in June 1988.[4] All OECD nations are now beginning to wrestle with the challenge posed by its implementation. Some see the objective as a compromise between the natural environment and the pursuit of economic growth.[5] Some do not. Some recognise that there are structural as well as natural limits to sustainability, and agree that principles or criteria of sustainable development are necessary so that performance can be measured with the help of agreed bench-marks. Therefore, it seems urgent to work out such criteria, without which patchy performance and illusory progress might well result.

What follows is an outline of certain political factors and attitudes which may have an effect on how individual governments respond to the issues raised by the WCED. Of course, factors vary from country to country, combine in different ways, and have different weights and meanings, even in neighbouring nations. Nevertheless, it is likely that the OECD nations will be faced with most of these considerations when attempting to implant the concept of sustainable development into their respective societies.

FACTORS AFFECTING GOVERNMENT RESPONSES

Short-term Versus Long-term Perspective: A Matter of Vision

Because of the electoral timetable, the political mind-set is focussed on the short term. This leads to the formulation of policies which react to, rather than anticipate, crises. However, the nature of the issues raised in

Our Common Future requires that policy- and decision-makers shift to anticipatory decisions and concentrate on the long term. It may also require – as in Canada's case, with its abundance of fossil fuel resources – decisions which run contrary to strong and well-established economic interests and goals.

Shifting towards the long term might be easier in OECD countries where the WCED held public hearings; where former Commissioners are now promoting the Report; where the media give sustainable development extensive coverage and keep the flame burning; where the scientific community is active and engaged in an open dialogue; and where non-governmental organisations (NGOs) generate public support and highlight critical issues.

In the absence of the above, a promotional and educational programme may have to be initiated by the government. We know from experience that when governments choose an open consultation progress, the ensuing examination and discussion of issues generally produces positive results. When the public is involved, educated and informed, the quality of the policies chosen is better and their chances of implementation are considerably improved. The nature of global environmental issues offers an ideal opportunity for public consultations. Decision-makers may be able to influence public attitudes and the degree of acceptance of a required change: to succeed, decisions affecting consumers' and suppliers' behaviour in a sustainable economy require the informed consent of all parties.

Assessing Direct Versus Indirect Costs

When an environmental initiative reaches the political level, the first hurdle to be cleared is the financial one. Most significant initiatives require a substantial commitment of funds, often at a time when revenues are shrinking. In addition, opponents are quick to circulate data on the so-called economic cost of the proposed action, usually in terms of jobs lost. Proponents are left to search desperately for an assessment of the true cost to the economy of not taking action. Data are often not available and such costs are not easily calculated. If sustainable development is to become a political reality, credible empirical data on the cost of inaction must be secured and made readily available,[6] and conflicts between environmental and economy policy must be reviewed and resolved.[7]

The Human Impact

Governments accurately perceive that the sustainable management of natural resources could lead, at least temporarily, to lower rates of harvesting, the closure of some single-industry towns and to other negative social effects. The transition from an unsustainable to a sustainable economy requires adjustment policies, social 'shock absorbers' and thoughtful planning, all of which may be costly initially, and socially disruptive in the short term. So far, none of the OECD countries appears to have developed a policy for the social consequences of implementing sustainable development.

The implications of not acting will also be serious, however. For example, communities in Canada which rely on fisheries will soon require appropriate measures to compensate for a rapidly diminishing resource. The management of social disruption brought about not only by depleted fisheries, but also by desertification, deforestation and other causes, will inevitably need to be addressed. The costs of this are likely to be higher than those of anticipatory action.

Competitiveness: At What Cost?

Sustainable development can be interpreted as a damper on growth and economic activity as known today in the world's consuming nations. Some OECD nations generously endowed with natural resources, Canada among them, are clinging to the belief that low prices of energy and raw materials are the answer to achieving a competitive edge. Thus they engage in industrial activities that cause severe pollution and follow unsustainable energy and resource paths, out of fear of losing their competitive advantage. However, if, as the WCED proposes, sustainable development could be understood to mean a more efficient use of energy and resources, which in turn should lead to enhanced competitiveness, then a positive attitude to the concept would prevail.

The Market Place: Far From Being Free

William Ruckelshaus, a former WCED commissioner (USA), writes: 'The question then is whether the industrial democracies will be able to overcome political constraints on bending the market system toward long-term sustainability'.[8] A neat phrase which reveals, by inference, that the market place is rigged in favour of short-term gains and that many governments are loathe to intervene in the economy.

Distorted, as they are, by fiscal and taxation measures, and by the world-wide proclivity to externalise the costs of production as much as possible, the forces at play in the market place deserve close scrutiny. Factors working against efforts to internalise the costs of production include fear of losing markets to competitors who are not subject to a comparable regulatory regime; fear of changes in consumer behaviour; lack of capital and incentives for the introduction of appropriate technology and equipment; and price distortions (of water, energy, raw materials, etc.), which send the wrong signals to consumers. A new generation of incentives, regulations and penalties will need to be introduced in order to 'bend' the market place towards long-term sustainability.

Institutional Inertia: Between Rhetoric and Action

Much has been said about the institutional and bureaucratic inertia which is an important impediment to the implementation of the WCED's recommendations. In many instances, however, the public and NGOs have provided an impetus that has helped to shake up political bodies and overcome institutional inertia. An informed public and mobilised public opinion can make a substantial difference in determining the behaviour of governments in response to sustainable development. In some OECD countries the public has a remarkably high level of understanding of issues and has expressed its readiness to pay for the necessary action, as well as to change consumer behaviour in other ways. The media have sensed this interest on the part of the public and have played a major role in influencing politicians and decision-makers. However, institutional inertia is also a symptom of other factors, such as a serious lack of political leadership and determination in the major OECD countries.

In most OECD countries the political response to sustainable development has had a high profile, with frequent declarations of commitment. The latter make good television clips and headlines. When it comes to application, it is the Minister for the Environment who is left to carry out the task while cabinet colleagues raise doubts as to what the public is willing to accept: 'There may be short-term hardships!', 'There may be an impact on growth rates!', 'The unknown political consequences may cause problems!'. In short, politicians do not like uncertainty. Until the public expectations are understood by politicians, political rhetoric could continue to overwhelm political action.

THE CANADIAN APPROACH

By comparison with other countries, an unusually large proportion of Canadians are aware of the WCED's Report. Indeed, it is almost a best-seller. It has been translated into French and has become obligatory reading for those interested in global issues. Since the WCED published its Report, the media have shown considerable interest in sustainable development and have informed the public with special feature articles and editorials. NGOs have created innumerable themes inspired by sustainable development and have vigorously promoted its principles.[9] The scientific and academic communities have also shown a strong interest. For example, the Royal Society of Canada and the Institute for Research on Public Policy (IRPP) held a conference on 'The Brundtland Challenge' in March 1988. The IRPP created a Sustainable Development Programme under WCED Secretary-General Jim MacNeill. A one-week public event in Toronto devoted to the application of the principles advanced by the WCED was organised by a Member of Parliament. The University of Western Ontario held a one-day symposium devoted to an examination of sustainable development. Innumerable other initiatives have taken place, organised by institutes, NGOs, government agencies and the private sector, for the purpose of discussing the application of the WCED Report. Indeed, in October 1986, in response to the WCED's visit to Canada and before *Our Common Future* was published, the Canadian Council of Resource and Environment Ministers, composed of provincial and federal ministers, appointed a National Task Force on Environment and Economy for the purpose of bringing businessmen, government officials, academics and representatives of the public together to examine the interrelationship between the environment and the economy and ways of integrating the two in the decision-making process.[10]

In Parliament, an Opposition Day Motion endorsing sustainable development was debated on 15 May 1987.[11] In 1989, the concept of sustainable development was included in a Government Bill creating the Department of Forestry by way of an amendment proposed by the Opposition.[12] A similar amendment by the Liberal Opposition to the Bill creating the Department of Industry, Science and Technology was rejected by the Government.[12] The Standing Committee on the Environment is currently studying the implications of sustainable development and global warming. The Prime Minister of Canada announced in a speech given to the UN General Assembly in 1988, and has confirmed several times since then, his intention to open in Winnipeg an international centre for the promotion of sustainable development.

The responsibility to guide the government of Canada in response to the WCED's report was given to the federal Department of the Environment (Environment Canada), where a Sustainable Development Branch was created to 'identify the range of economic and fiscal instruments available to us and to examine potential ways of modifying them in support of environmental objectives'.[14] In August 1989, Environment Canada published a list of policies that should be modified so as to render them consistent with environmentally-sustainable development. They fall into seven broad categories:[15] regulations; fiscal incentives; sectoral policies; regional development incentives; management policies for real property; operational guidelines; and research and information activities.

In addition, Environment Canada proposed changes in the wording of economic regional development agreements to reflect broader consideration of the environmental opportunities and constraints in preparation of economic initiatives.[16] As Environment Canada cannot unilaterally modify existing policies, action will be required by the respective departments. In 1988, Environment Canada set up an independent panel to identify and promote products determined to be 'friendly' to the environment, to help the consumer make informed choices about the products in the market place.

A number of other initiatives have seen the light of day in Canada since the publication of the WCED's Report. Amongst them are: the inclusion by all three national political parties of sustainable development in their platform for the November 1988 election: the endorsement in June 1988 by Canada, as a member of the G-7 group, of the concept of sustainable development; the formation of one national and ten provincial Round Tables on the environment and the economy including the federal ministers responsible for the environment, finance, and science and technology; the inclusion of the Minister of the Environment on the powerful Planning and Priorities Committee of the federal Cabinet; the inclusion of the environment and sustainable development as a separate item on the agenda of the First Ministers' Conference in November 1989;[17] a commitment by the Federal and Provincial Fisheries Ministers to implement a sustainable fisheries policy;[18] a policy statement by the federal Department of Agriculture that refers to sustainable agriculture;[19] a commitment by the Canadian International Development Agency to the implementation of an environmental policy for foreign aid initiatives;[20] a study by the Canadian Environmental Advisory Council on sustainable development in Canada;[21] a study by the Science Council of Canada on the research needs for the development and implementation of sustainable policies in Canada;[21] and finally the participation by the Prime Minister in the Hague meeting of 11 March 1989, and Canada's signature of the Hague Declaration.

These are process-orientated initiatives, policy recommendations and declarations of intent. It is hard to know at this point whether these initiatives are a prelude to policy decisions of substance or will remain what sceptics consider to be merely good public relations gestures intended to create a positive impression, to reassure the public that government is taking the WCED's proposals seriously.

So far, the absence of a substantive domestic policy to implement sustainable development raises a number of questions, in Canada as in other countries. For example, why, despite the commitments to sustainable development and a new environment-economy ethic, does Environment Canada's budget rank fifteenth of all federal departments, and comprise less than 1 per cent of the government's total budget?[22] Why has Environment Canada's budget lost more than 70 million dollars to inflation while other departments, including National Defence, have not only kept up with inflation, but have also received real budget increases? Why has the presence of the Minister of the Environment on the Priorities and Planning Committee of Cabinet not produced any tangible action towards the integration of environmental objectives in key federal departments such as energy, transport, agriculture and finance?

Why do the 1989–90 spending estimates of the federal Department of Energy, Mines and Resources increase Canada's dependence on the exploitation of fossil fuels, while progress to sustain and expand renewable sources of energy is being phased out?[24] Why does the federal Minister of Energy promote an enhanced supply of energy, primarily fossil fuels and nuclear power, instead of promoting policies that aim to conserve and manage the demand for energy? Why have deficit reduction measures been used as an excuse to cut research in renewable and alternative energy sources, while a commitment to spend over 5 billion dollars to develop fossil fuel megaprojects remains untouched? Why have none of the six federal-provincial forestry agreements which expired in March, 1989, including the key provinces of British Columbia, Ontario and Quebec, been renewed?[25] Why do recent tax reform proposals not include measures to encourage private sector intiatives for the protection of the environment, the development of desirable technologies, economic activities that are environmentally friendly? Why do government procurement policies not favour supplies and equipment that are environmentally friendly?

THE ROLE OF THE OECD

Understandably, no single nation can solve the environmental crisis alone and set sustainable development in action. The role of the OECD is becoming very important in helping its members bridge the gap between rhetoric and action and integrate economic and environmental considerations. Since November 1988, its programmes have been restructured to a considerable extent to incorporate environmental concerns in its work.[26] It is likely that the 1990 OECD programme will define the environment as a horizontal issue, requiring attention across the organisation. Also, the role of aid in helping developing countries address their environmental problems is likely to be given high priority.

OECD ministers have reaffirmed the critical importance of integrating environment and economic decision-making. The OECD will work to place environmental decision-making on firm analytical grounds, with particular attention paid to breaking new ground in integrating environmental considerations into economic growth models, analysing environment-trade relationships, determining how price and other mechanisms can be used to achieve environmental objectives and elaborating the 'sustainable development' concept in economic terms.

The OECD will examine incentives and barriers to the innovation and diffusion of environmental technologies. It intends to seek coordination of policies among member countries with a view to promoting mechanisms for technology transfer to developing countries. It also plans to coordinate policies aimed at: balancing long-term environmental costs and benefits against short-term growth objectives; designing innovative approaches by development assistance institutions to environmental protection and natural resources management; and integrating environmental considerations into development programmes, taking into account the legitimate interests and needs of developing countries in sustaining the growth of their economies and the financial and technological requirements to meet environmental challenges.

The OECD Industry Committee is now analysing the effects of environment-related measures on industrial investment and performances. The committee will also examine policies which have been implemented in response to the impact of environmental problems on industrial adjustment.

The OECD Environment Committee will make an analysis of the net impact of environmental regulations on industrial productivity, technological change, profitability and mobility. It will also investigate ways to develop and implement environmental regulations through new forms of cooperation between government and industry.

Finally, the OECD Development Assistance Committee and the Development Cooperation Directorate will assess how bilateral and international assistance can contribute to ensuring that environmental concerns are adequately addressed in the design of aid policies, programmes and projects; to strengthening developing countries to improve their environmental policies and programmes; to examining the role of aid agencies in dealing with global environmental issues; and to aiding in the prevention of environmental disasters.

THE POLITICS OF LANGUAGE AND THE LANGUAGE OF POLITICS

Despite the conceptual progress being made by the officials at the OECD, the representatives of governments participating in the Meeting on the Environment of the Conference on Security and Cooperation in Europe (CSCE) revealed a deeply-ingrained preference for the phrase 'environmental protection'.[27] Rarely was the term 'sustainable development' used. When it was used, it was within a context of environmental protection. The strong attachment to the language of environmental protection shows that most governments and decision-makers still see the environment and the economy as *separate* entities. This leads to the conclusion that – three years after the publication of *Our Common Future* – the conceptual *separation* between the environment and the economy still prevails. For OECD governments, the environment has yet to enter the economic debate and will remain out in the cold so long as economic goals are seen as separate from environmental goals. With the focus on environmental protection, OECD governments – when facing difficult economic times – will set aside environmental goals, perceiving them as obstacles in the pursuit of immediate economic objectives such as employment, growth and GNP.

Yet in increasing numbers these same decision-makers refer to the term 'sustainable development' in their public statements. But what do they mean by that term? Integrating the economy with the environment? Living off the interest generated by natural resources? Efficient use of energy? Internalising the costs of production and pursuing zero discharge goals? Encouraging waste reduction, recycling, re-using? All of the above? Some of the above? Has sustainable development entered the political lexicon for a while, to be discarded once out of fashion? Only time can tell.

Governments burdened with debt due to large military expenditures and that are preoccupied with deficit reduction will find it extremely difficult to integrate the economy and the environment. To introduce clean technology, modernise dirty industries, build municipal infrastructure, put incentives in place so that air, water and soil no longer become burdened with production costs which are presently shifted from the factory to the environment, will require a considerable shift in political priorities.

But the political mind has not yet reached that threshold. This conclusion is prompted by the following observations:

- as recently as October 1989, thirty-five governments (including twenty-one OECD members) gathered at the CSCE Meeting on the Environment in Sofia could not agree to adopt the user-pay principle when deciding on ways of dealing with pollution in international waterways;
- most OECD member countries seem incapable of implementing the polluter-pays principle;
- most OECD governments, except perhaps for the Netherlands, have yet to tackle the broad range of issues at the heart of sustainable development;[28]
- several OECD nationals persistently overfish in international waters;
- Iceland is the only OECD nation to ratify the Law of the Sea Treaty;
- the United States and Japan refused to join seventy nations in endorsing a commitment to control the emission of carbon dioxide by the year 2000, citing a need for 'further study'.[29]

Crossing the political threshold requires interventions in the marketplace, and the introduction and enforcement of regulations. This poses difficulties to conservative governments, as such measures are not compatible with their ideology.[30] But it may be that all traditional parties, having for so long endorsed traditional economics, short-term interests and rampant consumerism, are incapable of making the conceptual somersault required if we are to integrate economic and environmental policies, see the environment as the sustaining base of civilisation, shift budget priorities, internalise costs of production, and eradicate poverty from the world scene. Motivated by increasing public anxiety and by the urgency of the situation, political parties will have to decide whether to become the vehicle for the sustainable development agenda, or irrelevant. But is this problem limited to political parties only?

Deep down, at the root of these important decisions looms a broader societal question: are we ready to live up to the growing conviction that the well-being of humankind – be it in economic, health or security terms – depends on how we behave on this planet, manage its resources, and redistribute the wealth we generate? If we are, we can make the required changes in the way we manage our affairs. And we must, if we are to look with confidence into the future.

12 Common Future – Common Challenge: Aid Policy and the Environment

Rt. Hon. Christopher Patten PC, MP

Groucho Marx was once asked a simple question to which he did not know the answer. 'But every school boy knows that,' he was told. 'Well,' he replied, 'go outside and get me a school boy.' It is obvious from my own breakfast table that schoolgirls are at least as well informed as schoolboys. My youngest, nine-year-old, daughter lectures me over the marmalade on the ozone layer, acid rain and the destruction of tropical rainforests. I say nothing of whales! You would have to be a pretty bull-headed father, or an extremely obtuse politician, not to recognise that the world has turned. Environmental issues are no longer confined to the bottom of the inside pages of serious newspapers. Instead they are front page news and prime-time television. Environmental issues are moving fast up the political agenda. I hope the debate will be increasingly well-informed.

When Prime Minister Brundtland launched *Our Common Future* in the Queen Elizabeth II Conference Centre in April 1987, she knew that a London launch would maximise world-wide publicity. She even made sure she had those essential modern marketing tools – an excellent video and a couple of TV series. However, the World Commission's report did not really capture popular imagination and concern. As far as I am aware *Our Common Future* never entered any 'best seller' list. The Government's reply, *Our Common Future – a UK Perspective*, barely made the newspapers at all, even though it was and remains one of the few national responses to the Report. *Britain and the Brundtland Report*, a booklet by seven organisations ranging from Oxfam, through the World Wide Fund for Nature, to Friends of the Earth, also sank virtually without trace – something I particularly regret since it complimented me and my then Permanent Secretary (who actually deserved the compliments) at least four times! Six of those organisations have just published a critique of the government's response and I look forward to discussing the aid issues in their paper with them.

No, the Brundtland Report may have articulated concerns which are of increasing importance to our age, but I fear the real impetus for their rising political profile was a series of natural, man-enhanced and man-made disasters. The Report itself recorded that during the 900 days the World Commission was at work: the African drought put 35 million people at risk and may have killed a million; the Bhopal accident killed 2,000 and injured 200,000 more; the Chernobyl reactor explosion caused damage across Europe; a chemical fire in Switzerland poisoned the Rhine, all the way to the Netherlands; and least heralded and most shocking of all, 60 million people (more than the population of Britain) died of diarrhoeal diseases caused by malnutrition and dirty water.

Since then we have had floods in Bangladesh and Sudan, Hurricane 'Gilbert', and the earthquake in Armenia. In all four cases, man's failure to respect his environment (manifested by deforestation, by his ripping out of mangroves and hence coastal defences and by his failure to enforce building codes) added to the scale of destruction. It seems that we still find it hard to give serious attention to major problems until we are forced to do so. You could be forgiven for thinking we are no wiser than when A.E. Housman wrote:

> The signal fires of warning
> They blaze but none regard
> And on through night to morning
> The World runs ruinward.[1]

But the case is not hopeless. Ruin can be averted. Environmental issues are coming sharply into political focus. As I shall show, action is under way in a variety of fora.

I should like to start by drawing two distinctions. First, we must distinguish between the different types of impact on developed and developing countries and be aware that solutions in the two pose different degrees of difficulty. The problems faced by developing countries are often more acute; their environments are more fragile and more liable to flood, cyclone and drought.

In saying this, I am not minimising our own domestic environmental challenges and opportunities. Developed countries have made mistakes in the past which will prove costly to correct. Ensuring that our future development takes careful account of the environment, and that it i based on sustainable growth, will involve dislocations and economic trade-offs. It will need imagination to make full use of the new possibilities available to us. But Britain and other donor countries are rich, with good soils, temperate climates, educated, healthy and well-fed people, long scientific traditions and democratic institutions. We face the problems of affluence and have the resources to determine the quality of growth. We have the luxury of making choices, and probably both the time and the resilient resource base to be able to afford some mistakes. We exercise those choices in an atmosphere of vigorous debate.

Poor countries have to deal with the problems posed by fragile ecologies and uncertain weather against a background of poverty, rapidly expanding populations, lack of qualified manpower and a limited scientific base. While they face exceptionally difficult choices, many countries do not have the democratic institutions needed to develop a consensus which all can respect. That said, developing countries perhaps stand to gain even more than we do from responsible use of the environment – and at least they may be able to avoid the mistakes we have made. Experience has taught us that prevention is cheaper than cure. If one addresses environmental concerns early enough in implementing projects, a 'green' approach may actually save money.

My second distinction is between two classes of environmental problems – the global and the local. Global problems are those that inevitably affect the whole of mankind and can only be solved if all the world's nations collaborate. Local problems are to some extent 'location specific' and are amenable to national or regional action. Of course, in the real world the distinction is not clear-cut. So, for example, widespread deforestation is global to the extent that it destroys species and increases carbon dioxide, but also local in its effects on soil erosion and rainfall. Most people would surely agree that global problems include depletion of the stratospheric ozone, the build-up of greenhouse gases, climate change, loss of genetic diversity and the exhaustion of major irreplaceable natural assets. In a sense it is the collective scale of national or regional problems which produces the global threats.

The local problems of principal concern to developed countries are largely caused by reckless industrialisation and unthinking affluence. They include the pollution of air, water and soil. This may be very widespread but it is still fit for solution by the North alone. Developing countries, on the other hand, face a long litany of local resource degradation problems including deforestation, desertification, soil erosion and salination. They suffer industrial pollution and the effects of rapid urbanisation, unclean water and inadequate sewerage disposal. One of the most important international environment issues we will face over the coming years will be to engage the developing world in the drive to tackle global issues; this will certainly involve helping them tackle what they see as their own priority environmental difficulties which may not currently pose any global threat.

In focusing on the international dimension I have been greatly helped by the excellent chapter by Sir Sonny Ramphal. He sketched out the scale of degradation through deforestation, desertification and pollution of soil, water and air. He graphically demonstrated what he called:

> 'The simple, and terrible, truth . . . that poverty and environment are inextricably interlinked in a chain of cause and effect'.

I share many of Sir Sonny's views on the environment, above all his stress on the links between poverty, population pressure and environmental degradation. I want to try to develop his arguments further and to suggest that there are practical measures we could and should be taking. There are many important issues such as global warming and the depletion of the ozone layer, but since I am Minister for Overseas Development, I propose to focus on two problems: that of population growth and the fate of the tropical forests.

Sir Sonny made the population issue more immediate by quoting statistics from just two countries. We need to look at the overall picture as well. The Brundtland Report cites United Nations figures which suggest that world population will stabilise somewhere between 8 and 14 billion or from 60 per cent to nearly 200 per cent above today's level. The range in these projections is enormous. It is as large as this precisely because we have the chance to influence the outcome. At the bottom end of the range we could increase food production per head, through foreseeable improvements in agriculture, and could keep present food habits. At the top end, we could not. There would have to be massive changes in diet and in our use of the environment.

It is one of the most important paradoxes of development that, in order to ensure that population growth stabilises near the bottom end of the projected range, we need to improve children's chances of survival. Provide mothers with literacy, education, employment opportunities, and child health programmes, and you provide alternatives to numerous births as a way of ensuring security in old age.

As Minister for Overseas Development, I try to explain that paradox to the public at large. Too many people in the rich countries still think that the poor of the developing world are feckless because the rational behaviour of the individual does not coincide with the good of the local or the global community. The reason why very poor families have so many children is because so many of their children still die before reaching puberty. If development efforts and aid programmes have one fundamental purpose, it is to stop so many babies dying. That does not just need primary health care programmes, but sustained increases in incomes and improved sanitation, agriculture, education and the rest. One good way to measure whether real progress is being achieved in a developing country is to look at the infant mortality figures.

It is the same with environmental degradation. Poor people do not deplete their resource base out of perversity. As experienced farmers, the rural populations of developing countries know only too well the consequences of over-using marginal lands. But they do not have resources on which to fall back in hard times. Their savings are not put into cars or domestic luxuries; they take the form of trees or cattle. Once those are gone, survival itself dictates the choices. Help a family today to find a way out of dire poverty and sustainable use of resources tomorrow becomes a possibility. Poverty and population pressures, which are increasing faster than people's ability to adapt systems of production, are very powerful causes of environmental degradation. It is no wonder that the symptoms of that degradation are so alarming.

Sir Sonny properly drew attention to tropical deforestation and quoted the FAO figure of an annual loss of 11 million hectares. That figure represents an area the size of Scotland, Wales and Northern Ireland. It is the most widely quoted because it is the latest we have for global forest loss drawn up from internationally comparable data. However, it relates to 1980 and certainly under-estimates the present position. Figures for the burning season of 1988 produced by the Brazilian Space Research Institute suggest that Brazil lost 30 million hectares of savanna woodland and rainforest in that year alone. British newspapers under 'shock horror' headlines described that as being an area equivalent to the size of Belgium. The reality is even more horrifying. Thirty million hectares is ten times as big as Belgium.

Why does such destruction matter? After all, deforestation was so widespread in Russia and North America in the 1860s that it was then responsible for the largest release of carbon dioxide into the atmosphere.

The expansion of agriculture, which followed this burning and cutting down of the forests, provided the basis for subsequent industrialisation and growth.

Unfortunately, today's burning of the rainforest is unlikely to have such providential results. The soils below the lush vegetation of Amazonia are fragile and poor. When the forest is cut down and burnt, the ash gives a temporary boost to the soils, but the minerals are soon leached out and the organic matter, with its valuable water-holding and soil-conditioning properties, is destroyed. The poor migrants from the drought-ridden north-east of Brazil or the slums of its large cities all too often find that their dreams of growing abundant crops on a secure homestead are an illusion. Their back-breaking work has merely cleared land for absentee ranchers. Even pasture deteriorates and ten years after clearing, much of the land is hard-baked desert or scrub unpalatable to cattle.

This cycle is tragic enough, but in the process a resource of global environmental importance is being destroyed. Forests, and especially rainforests, can play a major part in checking the gradual warming of the earth's atmosphere. Mankind releases carbon dioxide into the air from power stations, transport systems, factories and homes. While they grow, the forests counteract this, in part, by absorbing carbon dioxide and so help to stabilise the earth's climate. But as long as they are being cut down and burnt at present rates, this contributes between 10 and 20 per cent to net global carbon dioxide emissions. So, while power generated from fossil fuels (and most such power is generated in rich countries) is by far the most significant source of greenhouse gases, we cannot afford to ignore tropical deforestation in our studies of the agents of climate change.

In a global context, the rainforests are important for their genetic diversity. Because almost all the insects have yet to be identified, no one knows whether rainforests harbour two or 50 million species! We do know that the complex ecosystem of the forests means that many of the species are highly adapted to their ecological niche and have unusual and therefore valuable properties. It is no accident that 40 per cent of all drugs prescribed in the US are based on chemicals derived from rainforest species. Nor is it an accident that the US National Cancer Institute has identified 2000 rainforest plants with the potential to fight cancer.

Darwin taught us that species loss is an inevitable part of natural selection. Probably 98 per cent of all the species that have existed over the past 400 million years are now extinct. That works out at one loss every twenty-seven years. Burning of the forest for unsustainable agriculture means we are losing species at a rate of at least two a day. You may recall that Lady Bracknell told Ernest, 'To lose one parent may be regarded as a misfortune; to lose both looks like carelessness'.[2] If the World Resources Institute is right and we lose 13,000 plant species by the year 2000 from Latin American rainforests alone, our wanton destruction will have been more than careless, and our descendants will talk of calamity, not misfortune.

The British public is most concerned about loss of rainforests because of these global consequences. But as an aid donor concentrating on the poorest countries, I am at least as worried by the damage to the local environment by loss of forests throughout the developing world. Deforestation affects regional weather patterns and rainfall: there is evidence of this from Amazonia and from West and Central Africa. The forests of Rwanda need to be protected, not only for the sake of the mountain gorilla, but also to protect the livelihood of Rwandan farmers. Trees are of the greatest importance in arid areas – a fact that was once described by a Somali pastoralist thus:

> Villagers have no factories or industries. We have land, we have water and after that we need trees. Allah created trees because animals need them just as people need land . . . our lives depend on our animals and they depend on trees. The trees save our lives in drought. Trees are male camels for us – they bear our burdens.[3]

In the West, 'energy crisis' means sudden shocks in the price of oil. In Africa it means shortage of fuelwood, the price of which, assuming it is available at all, is measured in women's time. As wood becomes even more scarce, dung is burnt instead. That means essential nutrients are no longer available for crops. The crops foregone each year because of that loss of natural manure may well be greater than the annual total food aid provided by all the world's donors.

Moreover, organic matter and trees both help to prevent soil erosion. The World Bank estimates that to make up the nutrients carried off by soil erosion would cost India $6 billion in replacement fertiliser each year. That is forty-five times our substantial aid programme to India. Soil degradation wreaks appalling human misery. The Horn of Africa has had more than its share of suffering in the 1980s. Displaced people trek back and forth across hostile land. We are faced with the heartrending sight of starving babies in feeding camps – that appalling testimony to the failure of every sort of development, political as well as economic. The immediate causes are war and drought, but these people are also environmental refugees. It is no coincidence that Ethiopia's descent into acute vulnerability to famine follows the stripping of forest cover from its Highlands. Ethiopian forests now cover one-tenth of the area they did in 1900. Lands stripped of their tree cover cannot retain what moisture there is. When rain comes, it is not absorbed. We are faced with the mocking spectacle of disastrous floods in areas usually regarded as arid wastes. When life is precarious, people are more likely to fight over precious water and pockets of good soil. Tribal loyalties become more important if group survival is in question.

So one type of degradation alone, deforestation, can have profound consequences: loss of genetic diversity, climate change, loss of soil fertility, soil erosion, flood and famine. And I should stress again that deforestation is only one example of the environmental problems we need to overcome.

This may all have seemed a touch apocalyptic, and we must not be paralysed by pessimism. What we need is to be imaginatively, practically and resiliently constructive. Indeed, that is the virtue of the Brundtland Report. The earlier Club of Rome Report was politically, not to say economically and technologically, naive in insisting that we must all give up the benefits of twentieth-century affluence. By preaching 'no growth', they ensured they had 'no influence'.

The World Commission, chaired by a serving prime minister, and including eight ministers or former ministers, was much more astute and realistic. The Report does not say 'Carry on as you were', but it does give a message of hope. It points to sustainable growth and the elimination of poverty as the basis for an environmentally sound world. It asserts that technologies can be found if the will exists. Faced with that approach, which politician would dare to cast doubt or refuse to take up the challenge?

I believe profoundly in the importance and possibility of sustainable development. It is, of course, difficult to define simply, but it was neatly characterised by the Brundtland Report as meeting the needs of the present without compromising the ability of future generations to meet their own needs. To bring that about requires three things: political will, resources and the right mechanisms.

Political will appears to be available in developed countries and in some developing ones. The Prime Minister's speech to the Royal Society in September 1988 made her personal commitment quite clear and showed that growth and environmentalism can go together. She pledged that: 'The Government espouses the concept of sustainable economic development. Stable prosperity can be achieved throughout the world provided the environment is nurtured and safeguarded.' President Bush stressed environmental issues in his election campaign and Budget Address, and has nominated the President of the American World Wildlife Fund as head of the US Environmental Protection Agency. President Gorbachev, who has an agricultural background, apparently has plans to double expenditure on environmental protection by the early 1990s. President Mugabe and Prime Minister Gandhi both chose to take part in the UN debate on the Brundtland Report.

Two swallows do not, alas, make a summer. Other developing countries will need to follow the lead taken by Zimbabwe and India. They will also need to join in action on both global and local problems. As I have argued already, that is more difficult for developing countries than it is for us. We can and we must help to generate the necessary political will.

Generating political will involves action and persuasion at several levels. First, to have any credibility at all, we in the West must solve our own problems. I call this the older greenhouse effect – people who live in glasshouses shouldn't throw stones! Developed countries must adopt the right policies to deal with both global and local environmental problems. Only if we do that can we hope to convince developing countries that global issues, like climate warming and the ozone layer, affect them as much as they do industrial nations. The Maldives know that global warming and a rise in the level of the oceans could mean disaster for them. But other developing countries may regard global preoccupations as beyond their concern. They will point out that it is the North's demand for thermal energy, not the burning of forests, which is the main source of carbon dioxide emissions, and they will emphasise the enormous scope for energy-saving in the North. Developed countries will have to show that they are doing all they can to solve the problems they themselves are causing.

Second, we must make clear that we recognise that developing countries will have different priorities. Their own local environmental problems, of drought, flood control or soil erosion, will often be of greater urgency for them. They will insist that global concerns should not penalise them for their present poverty; they need the freedom to develop and industrialise. If energy-efficient and 'environment-friendly' industrialisation demand higher additional costs, developing countries will naturally push their own priorities. They will seek more open markets, so that they can earn the necessary resources to afford action, and they will be right to do so. The poorest countries will look for greater and more effective aid. The developed countries for their part will need to show appreciation for recipients' domestic concerns and priorities. There may be scope for packaging assistance to cover not only those factors which contribute to global problems, but also help for local environmental issues such as soil erosion or fuelwood shortage. Both partners (the aid donor and the recipient), and the world at large, would gain.

Some have tried to short-circuit this co-operative building of political will. They have argued for using blanket environmental conditionality against developing countries who seem to be ignoring the global environmental good. It has been suggested that aid should be conditional upon a developing country's policy on all environmental issues. As some may know I certainly do not reject the principle of conditionality linked to aid. But it must be sensible and practicable conditionality. I certainly believe that one should encourage environmental concern in aid recipient countries, for example, by trying to persuade them to give forestry projects a higher priority. This is an example of the policy dialogue that is such an important instrument for ensuring that our aid is effective. Equally, it is right to make specific project conditions about pollution standards. What I oppose is the argument which says one should cut off all aid to a country which does not immediately respond, for instance, to what the outside world says about its approach to this or that aspect of environmental policy. That smacks of sanctions. What we want are enthusiastic converts, not press-ganged laggards. If we do not help countries to develop their resources sustainably, and to find alternative livelihoods, it is all too likely that they will be forced to pillage their existing resources even faster.

Equally, I cannot agree with those North American pressure groups who seek to make progress by imposing their own domestic environmental protection legislation on development projects. For one thing it is bad ecology. Standards set for temperate, highly urbanised environments may make no sense in tropical rural areas. For another, it wrongly assumes that developing countries have the basic data and monitoring capacity that would be needed. While global standards are often inappropriate, internationally-agreed guidelines linked to local or regional factors may well have a role to play here. As part of a more constructive approach, there is the need to strengthen institutions in developing countries so they can decide on appropriate standards, which they can monitor and enforce. We are doing just that ourselves, for example, by financing a number of trainees in environmental disciplines. It might seem to take longer than conditionality, but the results will, we believe, be more sustainable.

Resources are not just a question of money, and it is important to remember the Brundtland Report's strong message that environmentally-sensitive development could often reduce costs. Bearing this in mind the first need is to follow the Prime Minister's dictum in her Royal Society speech: 'We must ensure that what we do is founded on good science to establish cause and effect.' I suspect that members of that Society would not be surprised to learn how many gaps there are in our understanding of tropical ecosystems and of the interactions of development and environment. To devise solutions which are simple, appropriate, affordable and therefore sustainable, we will often need the best scientific minds. A Nobel Prize has not gone to those who made the oral rehydration therapy salts, used to combat diarrhoea, cheap enough to be distributed on a massive scale. Yet the problem was not an easy one to solve and I suspect the benefit to mankind outweighs that from some much more prestigious scientific endeavours.

One area requiring attention is improving the genetic potential of trees which are hundreds of years behind agricultural crops in terms of selective breeding. We need to start by taking and labelling seeds of known provenance. To ensure the high status of such work we need wider recognition of its importance at the higher levels of academia.

One area where our understanding is particularly weak refers to the extent, expectant trends, and strategies for coping with desertification. A UN conference on desertification provided the opportunity for a baseline assessment in 1977 which was followed up by a questionnaire from the United Nations Environment Programme in 1984. The latter was based on a broader definition of the area at risk. Yet a comparison of the two has provided virtually all the widely quoted statistics on desertification used ever since. A whole United Nations Plan of Action to Combat Desertification was drawn up on the basis of the 1977 figures. The Plan calls for inputs of more than $4.5 billion each year but has never received backing on anything like that scale.

The Overseas Development Administration (ODA) has always preferred to tackle desertification as part of natural resource management rather than as a separate problem with its own priorities. The latest analysis from Ridley Nelson of the World Bank suggests we were right, and claims that there is little point in throwing huge sums to counter specifically what he prefers to call dryland degradation. The new study questions the scanty data and rejects the old concept of steadily advancing sand dunes moving southwards from the Sahara. Instead Nelson claims that, after allowing for fluctuations in rainfall, deterioration flows outward from centres of population pressure. So the problem may be as much socio-economic as climatic. The study urges research that is adaptable to local situations and rejects the famous grandiose plan to put a green belt across the Sahel from Dakar to Djibouti. Faced with such contradictory advice, a decision-maker could be forgiven for prevaricating and calling for more research!

The drive for improved science needs to involve scientists from developing countries, especially on work on global problems. We can expect the governments of developing countries to take more heed of predictions of global warming if their own researchers have played a part in formulating the assumptions or processing the results. A world scientific constituency needs to be built which can sensitise decision makers in North and South alike.

Apart from, or perhaps as a part of, good science, we need good economics. We need economics because the science will tell us what we need to do, while the economics should help to tell us how we can best do it. I make a fervent plea for economists and development specialists to step up work on environmental economics, for it is crucial that those worried about the environmental effects of projects and policies speak the language of economics if they are to influence ministries of planning and finance in developing countries.

Some used to object to cost-benefit analysis on the grounds that it did not take proper account of the long-term future, nor of factors which could not be quantified. Both points are important in assessing the environmental dimension of projects. However, we still need a mechanism for helping to allocate scarce capital amongst the developing world's possible projects. We commissioned one of our senior economists to write a practical handbook on integrating environment into project analysis. His text complements a general manual that we have produced for all our programme managers, which is intended to ensure that the environmental dimension is properly considered throughout the project cycle.

Apart from micro-economics, we are working with the World Bank at a 'macro' level. We have commissioned British consultants to study the trends in the supply of and demand for natural resources in the hills of Nepal, and to explore the possiblities for encouraging environmentally appropriate behaviour through macro-economic and other policies. The study is based on the World Bank's premise that degradation in developing countries is too widespread to be tackled by projects alone. The first phase suggested that on some reasonable assumptions food, fuelwood and fodder deficits would be common throughout the hills by 2011 and it might cost as much as $60 per head to import substitutes from outside. This figure is a third of current GNP per head. No wonder the Nepal National Planning Commission has shown great interest in what started off as a donor initiative, and has played a major role in selecting the policy instruments to be analysed in the second phase of the project. The World Bank is financing similar work in up to twenty-nine other countries. The series should provide some much-needed insights into the true causes of degradation.

Good science and good economics should help us make best use of the third kind of resource, money. They should also remind us that throwing money at a problem may not solve it. The most obvious way to promote sustainable development is to ensure that consumers and producers pay the full environmental cost of their activity. We all know countries where demand for electricity is almost infinite because tariffs are set very low or charges are not collected. The results are unnecessary thermal power stations and an environmental cost to all of us. It is not always easy to ensure that the market reflects full environmental cost, but we can make a start by eliminating harmful subsidies.

Those subsidies can take many forms. The European Community's Common Agricultural Policy and the agricultural support systems of the United States and Japan conspire to ensure that the world prices facing many developing country commodity exporters are far too low. That increases their need to deplete their resource base more quickly, and it reduces their ability to afford environmental protection. I am not saying higher prices would automatically improve the environment – they might instead lead to increasing pressure on more marginal land (as the CAP has done). My point is that more open agricultural trade is a necessary, though not a sufficient, condition for sustainability.

The argument is not just true for agriculture. Trade has a much bigger impact on most developing countries than aid ever can have or should have.

Many people would argue that we also need to think about debt in our effort to find resources for environmental protection. We have always rightly distinguished between two types of debt. The first type is that owed by the very poorest countries to governments and the international financial institutions – basically the debt of sub-Saharan Africa. The other type is that owed to commercial banks by the better-off developing countries, and here I am thinking primarily of Latin America. We launched in September 1988 our initiative on African debt and, thanks to the breakthrough achieved at the Toronto summit, Paris Club rescheduling is now under way, reflecting the Toronto consensus. Seven countries have already benefited. The problem of Latin American debt is again under active discussion. Our approach to these problems involves the avoidance of two particular risks. We do not want to transfer the problem from banks to taxpayers, nor do we want to deter the banks from lending to these countries again. Too many of the proposals made so far would have this effect.

Some environmentalists want to bring in additional resources through debt swaps. One might expect them to have been supported by northern pharmaceutical companies, who may have a more immediate interest in keeping rainforests than do hard-pressed governments of poor countries. Where such swops are entered into voluntarily, and genuinely release additional resources, they can be a useful tool, but I would not want to overstate their potential contribution. Naturally, if NGOs provide large sums of money for this purpose, they will need to be convinced that the environmental commitments undertaken are firm and durable.

As regards aid, I think that it is clear that official development assistance of the right quality, properly spent, is an essential part of the co-operative effort to find resources for sustainable development. Our own economy is now sufficiently strong for us to have started increasing our aid programme again in real terms. I hope that we can continue to afford to do that. We must also look for ways to focus our aid effectively, by means such as the forestry initiative which bore out the Prime Minister's pledge that the ODA would 'direct more of our aid to encourage the wise and sustainable use of forest resources'.

Let us now turn, finally, to the mechanisms for global sustainable development. By mechanisms, I mean the international institutions. For over forty years, international relations have been dominated by the consequences of an experiment that Rutherford conducted in Cambridge in 1919. I understand that when he split the atom he thought it would be of no practical use. However, nuclear weapons, the Cold War and disarmament have shaped alliances and preoccupied governments since the 1940s. I believe that, over the next forty years, the environment will come to form much of the business of international relations. Environmental protection, like arms control, will be complex and difficult to negotiate on an international scale. The two issues are alike, in that both concern the survival of the human race on this planet. With nuclear weapons, everyone realises that, however necessary they may be for security, they are extremely dangerous. Environmental damage, on the other hand, can arise from peaceful economic activities considered, until only very recently, as harmless, even beneficial. We are only beginning to understand the ecological principles at work, which have to be respected to ensure good environmental practice. All countries have a stake in the solutions. All need to be involved in finding and applying them. Inevitably the arguments about global environmental questions are bound to involve issues of sovereignty and equity, mutual interest and mutual help. It will not be easy to strike the right balance.

Emil Salim, who was a distinguished member of the Brundtland Commission, has described the rainforests as a global resource requiring and deserving finance from all nations for their protection. This is not the view of the Brundtland Report itself, which confined the global commons to Antarctica, the Oceans and Space.

A different sort of reaction comes, for example, from Brazil, which has been very jealous of its national sovereignty and territorial integrity ever since Spanish and Portuguese disputes over the Treaty of Tordesillas. In Brazil, rainforest has hitherto been regarded as a purely national resource to be used in accordance with national priorities. Interference by outsiders is deeply resented, although at the same time international opinion seems to be having an effect in raising the value of Amazonia in local perception. A local Brazilian senator reacted to a visit by five US senators and congressmen in January 1989 by writing in a Sao Paulo newspaper: 'If on the one hand the presence of these arrogant would-be lords of the earth tramples on our sensitivity, on the other it shames us that we have no strategy to exploit Amazonia in a way that preserves the environment'.[4]

No country, whether developed or developing, can be compelled to change its policies towards the environment. Any international system which sought to use compulsion would defeat its own object. Those who do not want to change their ways could easily evade its requirements. The only way to better environmental practice is to convince countries that it is in their own best interests to change. We have to show them that protecting the environment means a better life, both for themselves now and for their children. We have to make all countries understand that global environmental problems affect every one of them. There can be no free riders, for example, in the efforts to control climate change.

Developed and developing countries are already co-operating in identifying mutual interest in protecting the environment. But in the aid field, respect for national decisions has meant that concerned donors have so far provided money for conservation as part of traditional aid programmes. Projects have happened only when recipient governments attached priority to them; and it may be difficult for them to give increased priority to environmental projects when local people are crying out for schools or health care. But we in Britain have long recognised that improving living standards goes hand in hand with tackling environmental problems. I believe that this fundamental objective of our aid programme can only be achieved if recipient countries and donors work together on strategies for sustaining the use of renewable natural resources. Well-directed aid programmes can serve as an encouragement to the adoption of good environmental practice, as they are already encouraging sound economic policies in recipient countries.

Another powerful influence on the perception of common interest is clear, well-founded scientific evidence. The scientists have warned us plainly of the dangers of depleting stratospheric ozone by CFCs. On the basis of this evidence, an international regime for protecting the ozone layer has been agreed through persuasion and diplomatic negotiation. The Montreal Protocol marks a real breakthrough in international negotiation on environmental issues. But not all countries are involved. In particular China and India, which have the potential to be major producers of CFCs, are not signatories. In an effort to persuade all the non-signatories that CFCs pose a real threat to every nation, we hosted a major international conference on Saving the Ozone Layer in March 1989. This provided a showcase for new technology, for example in refrigeration, which will avoid the dangers of CFC emissions, and as a result more countries decided to join the Protocol. We have good arguments to convince other signatories, but in the end, no one can compel them to join.

Some might welcome a supra-national body with powers of compulsion. But to establish one would be a serious mistake, even more so because while we have the scientific evidence on CFCs, we do not yet have it on climate warming. We know the danger is there; it will be only too apparent if industrial emissions of carbon dioxide raise world temperatures and bring spring tides lapping over the Thames Barrier. At this stage, we cannot assess the scale, the timetable or the way different regions of the world will be affected. We need more science, and we need to prepare the economics, so that when the scientific results are in, we shall know the best measures to adopt, for developed and developing countries alike.

The perception of mutual interest develops slowly and pragmatically. Let no one kid themselves that we are ready for a culturally and climatically diverse global village to be ruled by an environmental policeman. The world's environmental plight is too important and too immediate for daydreams of supra-nationality to waylay us. We have no time for windy generalities and vapid pieties. Nor can we afford to waste years in wrangles about new institutions. We must use what we have. We must look out for and build on practical initiatives. It may not suit those in the green movement who want to overturn the world made by people, in order to preserve the natural world, but our approach must be heroically prosaic.

The successful work done to date on CFCs shows a way. The Tropical Forestry Action Panel with its emphasis on practical co-ordination and frank discussion is another model. The Inter-Governmental Group on Climate Change has made a successful start. The United Nations Environment Programme, fresh from its Montreal Protocol triumph, secured agreement on a Convention on the Transboundary Movement of Hazardous Waste.

All three initiatives share at least two features: a substantial British involvement and a connection with the United Nations system. The UN is all too often castigated as a talking-shop, but it is the most appropriate forum we have. If it did not exist already, we should need to invent it. But why on earth try to invent it twice over? The UN already has institutions dealing with the environment, social issues, refugees and agriculture. It is up to us to ensure that the UN functions as effectively as possible if we are to achieve sustainable development. That, naturally, begs a host of questions – but we would be better off trying to address those than engaging in the attempt to build new castles in the air.

Other institutions must also play key roles. I have mentioned the World Bank a great deal. Environmentalists, especially American ones, love to suggest that its only 'green' interest is the 'greenback'. It certainly helped to finance some projects which went disastrously wrong, as its greatly strengthened Environment Department would be the first to admit. But criticism tends to centre on six of the literally thousands of projects it has helped finance. I believe lessons have been learned. The Bank is now, as befits its stature, the leader in environmental research and policy initiatives amongst the donors.

I am also pleased that environmental collaboration between bilateral (country) donors is growing through the Development Assistance Committee of the Organisation for Economic Co-operation and Development (the OECD) in Paris. Environmental expertise is in short supply and we can usefully share experiences with each other. We also need to discuss together how to handle certain types of project. If global warming means that we should take more explicit account of greenhouse gas emissions from power stations in our economic appraisals, then we will obviously have to decide guidelines with other donors. Otherwise, if we suggest to a recipient that an energy efficiency project might make more sense, all that is likely to happen is that the recipient will take the efficiency project from us and ask another donor for the power station.

I depend for my conclusion, as I did for my initial remarks, on my youngest child. Seeking to explain to her the other day the meaning of Pandora's box, I looked it up in Lemprière's *Classical Dictionary* and read there the description of how Epimetheus opened the beautiful box, given to Pandora by Jupiter. When Epimetheus lifted the lid, 'there issued from it a multitude of evils and distempers which dispersed themselves all over the world and which, from that fatal moment, have never ceased to afflict the human race. Hope was the only one who remained at the bottom of the box . . .'

At the risk of abusing the fable, we don't want Hope to remain under wraps. The job for politicians in the next few months and years is to get Hope to fly. We should not delay. We shall of course save the earth. There is no other option. But we must start the hard, grinding, practical work now.

'Wise men', wrote Cavafy, wise himself, 'are aware of future things/ just about to happen.'[5] Even wiser men work together, when it is an implacable necessity, to stop some future things happening at all. We must and we will. 'Evils and distempers' may darken the heavens, but Hope is taking wing.

Part Five: The International Community and Sustainable Development

13 Environmental Advance and the European Community

Stanley Clinton Davis

There can be no doubt that the issue of environment and development is one of the most serious facing mankind, ranking alongside nuclear catastrophe in terms of its profound potential consequences for our children and those who will come after them. I believe that we must do everything we can to raise awareness of the appalling problems we face and of the pressing need to intensify our joint efforts to tackle them.

Increasing environmental damage and environmental hazards no longer affect only the industrialised countries. They are global problems besetting the developing countries as well. If we do not adopt radically different strategies to protect the global environment – if we do not comprehend that we live in *one* world – we can have no hope of safeguarding the environment on an enduring basis and we shall be threatened not simply by a series of crises but with catastrophe.

It was, I think, Indira Gandhi who described poverty as the worst form of pollution. What we must address is the combination of poverty and the degradation of natural resources which is the lot of so many of the human race, and which ultimately threatens the living standards and prospects of us all.

There will, indeed, come a point when the cries of the hungry will blow down the rich man's most firmly barred gate. That is why it is the height of hypocrisy for some developed countries to engage in a form of ecological colonialism, such as the entertainment of ideas, that all Third World aid should be tied to promises by recipients to protect their environment, when their own companies have so often played a major part in the environmental degradation of the developing world. That is why the European Community has for many years placed increasing emphasis on cooperating with its partners in confronting environmental problems under the successive Lome conventions. These conventions (I, II, III, IV) established agreements between EEC countries and former

European colonies in Africa and the Caribbean chiefly over trade preferences and the commitment of the EEC to buy produce.

In the third Lome convention there is a special chapter concerned with the fight against desertification. In 1986 the community plan of action for the conservation of natural resources and counter-desertification in Africa was adopted. Subsequently it is most encouraging that in the great majority of the indicative programmes under Lome III priority has been given to rural development and, within that context, to actions to protect the enviroment and natural resources and to the fight against desertification. Already numerous specific projects and programmes have been agreed. These actions, aimed at agricultural development and environmental stabilisation, are of the greatest importance in trying to reverse the steadily deteriorating ratio of mouths to feed, to hectares which can be cultivated.

The essence of the policies that the Community has been seeking to promote is to ensure that local people and local authorities are fully aware of the environmental problems that confront them and that they are active participants in seeking to remedy them. In this connection, no-one should overlook the vital role that women play in the life of these communities. They are – and must be entitled to – an increasingly important voice in determining the policies of salvation.

In the discussions which are currently proceeding with regard to Lome IV, the Commission of the European Communities has successfully urged that its negotiating mandate should include provision for the protection of the environment. We in the Community, and our trading partners too, must recognise that conservation cannot succeed without development. No-one can prevent debt-ridden Third World governments destroying their hardwoods in their quest for foreign exchange: the fact is that one rainforest tree can be worth as much as 1000 dollars in foreign exchange, and in supplying the market for panelling, parquet floors, luxury yachts and even good-quality coffins, the value can rise by seventeen times as much. Environmental pressures will only be reduced if Third World countries become less poor. The converse is also true. Development cannot be sustained without conservation. Thus governments and conservationists alike have a joint interest in finding solutions. Governments, unlike Lord Nelson, cannot continue to turn a blind eye to these massive problems.

Let me turn to European Community environment policy, which has evolved rapidly in recent years from being a fringe interest to being absolutely central to the construction of policies, both internal and external, upon which the future quality of life and prosperity of the

Community and, to an extent, the world, will depend. This is exemplified by the Single European Act, which came into operation last year and which amends the Treaty of Rome very significantly by adding a specific title on the environment. The three Articles (130R, S, T) of this title formally legitimise the legislation that the EEC has been developing since 1972 and advance the status and implementation of practicability of environmental policy in four ways.

First, they give symbolic importance to the environment. This is reinforced by the preamble to the Single European Act which refers to 'the fundamental rights recognised in the constitutions and laws of the Member States'. More than a third of Member States have accorded constitutional status to the protection of the environment or recognise constitutional rights. Second, they suggest a broad scope of objectives. For example, the proposed directive on freedom of environmental information is being based on these Articles. Third, the Articles give legal force to certain principles, including the most important principle that 'environmental protection requirements shall be a component of the community's other policies'. Finally, Article 130R lays down the principle to be used in deciding whether action is to be taken by the Community or Member States: can the objectives be obtained more successfully at the Community level or the level of the Member States? An implication of this is that all Member States need to be knowledgeable about environmental problems and policies in all parts of the European Community.

These concerns are, indeed, mirrored in the Community's fourth environmental action programme, which has been adopted in principle by the environment ministers of the Community. Regrettably, in too many instances, they seem to be departing from it in practice. It should be mentioned that, although the Community has no powers to enforce an action programme, such a programme may anticipate a forthcoming piece of legislation.

The consequences of desertification, deforestation, over-exploitation of the oceans and the other environmental challenges facing the Third World will affect literally billions of people. In the short term they pose the widespread threat of many miserable deaths. In the medium term they offer the unappetising prospect of widespread social and political disruption and the steady, often irretrievable, erosion of vital planetary resources.

What then of the long term? I believe that the answer to that question depends upon people like us all around the world. The responsibility rests on our shoulders. It is for us to bring about the change in attitudes and priorities which is vital if we are to avert the perils that face us and bring about the sustainable development for which the Brundtland Report so eloquently calls. Our planet is not simply an economic space – it is a *living* space, the protection and survival of which is largely dependent on international policy. Market forces need to be directed in favour of the environment – left to themselves they lead to disasters like Bhopal and Sandoz.

In times of acute economic difficulty, with the crushing pressures of indebtedness, there is a real risk of resorting to defensive, isolationist positions. The European Commission's belief has always been that suicidal 'beggar-thy-neighbour' policies can never regenerate hope for all our people and must, therefore, be replaced by 'better-thy-neighbour' policies. Hence the importance of the reform of the Common Agricultural Policy, which like the price supports and protection schemes of other richer nations, has driven down food prices with the dumping of surplus produce.

The Common Agricultural Policy has two main limbs – price policy and structure policy. Price policy sees to the fixing of commodity prices, and structure policy is concerned with measures designed to develop and modernise agriculture, improve productivity and strengthen codes of practice. The price policy was necessary when commodities were in short supply, but now only aggravates the effects of surplus. Fixed prices enable large farmers to make considerable profits, while smallholders merely subsist.

Moreover, structure policies have only been allocated a tiny share of the funds available when compared with the share given for price policy. Consequently, they have been ineffective in the very difficult task of closing the gap between standards of living and production of rich and poor regions in the Community. Average farm incomes in the Paris Basin region are now six times those in the poorest parts of Italy and Greece, and the gulf in the income between the richest and poorest farmers is far greater. In some countries, such as Italy, structure measures have failed to make much impact because national governments have not provided their share of the finance or have operated rival national schemes. As a result, the richer countries have often taken up a large share of the money available for Community-wide schemes. In 1985, the UK, with the largest farms, withdrew the most from Community funds under both the Less Favoured Areas Directive and the Farm Modernisation Directive.

Moreover, on the whole structure policies have been on too small a scale to have had much influence on the broad development of European agriculture, but have supported a variety of local schemes; some of them have also been environmentally damaging, mainly because of the insensitive use of subsidies: for example the land consolidation schemes of the kind common in the interior of Brittany in the late 1960s and early 1970s brought about the disappearance of hedges, banks, ancient trackways and ditches, and the flattening of the traditional landscape to create neo-openfields.

Such price supports have also led inexorably to the Third World being pushed further into subsistence farming with its hugely detrimental effects on the land and in the deepening of poverty. Part and parcel of all this is the imperative to strike a balance between the further development of industrial society and preserving an environment which is made all the more fragile by the pursuit of that objective. Economic and social development and environmental protection must be shown to be not only compatible but mutually reinforcing. We cannot afford to be pessimistic – although we must be conscious of the magnitude of the task that faces us – but I am conscious too of the resources of political imagination and initiative which are there to be drawn upon when necessary.

The European Community is itself an outstanding example of the possibilities for cooperation among disparate and, in the past, mutually hostile nations, when the political will is there. Recent examples in the environmental field – the ozone layer agreement, the international tropical timber agreement, the agreements that followed Chernobyl – illustrate vividly that the same will can be manifested in connection with global environmental problems. And it is vital for the future prosperity of mankind that the effort be devoted to the environmental problems of the developing world.

The European Community, particularly as a result of the Single European Act, is able to operate as an exemplary force for progress in the environmental field. The Treaty gives the Community the objective of improving the quality of life, and of the harmonious development of the economies of Member States, in order to promote steady and balanced growth. It has become clear that the regional management of environmental resources is essential to these aims. Even from a purely economic point of view environmental policy is a necessity. The Common Market would cease to function if Member States were to adopt differing national policies which resulted in disparities and distorted competition. Perhaps most important of all, pollution is rarely a national problem – it is no respecter of borders, as so many recent examples have clearly evidenced.

Within the Community, even before there was a formal legal base available, over 100 directives and regulations had been adopted covering all fields of environmental protection. There exist clear Community-wide rules on the testing and labelling of dangerous chemicals. There exist stringent standards to limit the likelihood of chemical accidents, and common standards for drinking water. The Community is taking steps to ensure the widespread availability of unleaded petrol. We have reached agreement to deal with atmospheric pollution from cars and other vehicles and from large combustion plants. One of the problems that we always face is that, in order to achieve agreement between twelve nations with different levels of industrial advance, with different problems, with different backgrounds and with different priorities, compromises are inevitable. This is bad news for some countries with high environmental traditions who dislike any compromise which might have the effect of achieving much less than they would think to be desirable. But, as Edmund Burke put it, 'Nobody made a greater mistake than he who did nothing because he could only do a little'.

Of course, new threats to the European environment continuously emerge and new inititatives – involving new negotiations – are constantly needed. Thus the fourth environmental action programme goes further than the previous programmes did by insisting that environmental considerations must be an essential integrated element of *all* economic and social policies. This is wholly in line with the Brundtland Report. There is no point in our trying to clean up water pollution if Community agricultural policy ensures a steady flow of nitrates into the continent's rivers, or in our backing the sort of industrial development in the south of Europe which produces smog of the kind we are busily cleaning up in the north.

There is an emphasis too on wider dissemination of information and increased public awareness – a move which was symbolised by the European Year of the Environment and which has been given additional importance by the draft directive dealing with greater access to environmental information, which I was able to persuade the last Commission to adopt near the end of its mandate. The basic rationale is that ordinary people are entitled to much more information about *their* environment and that too many organisations – government and corporate – deny them that information, often under the spurious pretext of 'confidentiality'. A well-informed public is vital in spurring governments to act to protect the environment. It will be interesting to see how governments will react to this democratic proposal. So far, regrettably, the auguries are not good.

Effective negotiations within the Community should result in stricter standards, which will not only increase environmental protection, but which will also encourage industry to recognise the need to respond to the growing demands for non-polluting products. This would result in increased competitiveness and employment opportunities, and aid the successful completion of the internal market by 1992.

Community environment policy must inevitably be outward as well as inward looking. The 350 or so million people of the Community exist in a world of almost 5 billion. Frequently the ultimate success of the work we do in Europe to achieve environmental improvements will depend on the success of similar policies adopted elsewhere. That represents another powerful justification for meaningful negotiations at all levels. For the most part, twelve states acting together in international organisations, such as the United Nations, can make a much greater impact than twelve acting separately. The Community has, therefore, rightly played an active part in international negotiations on environmental issues to date, either as a contracting party to specific agreements or within the framework of international organisations. In accordance with the Community's action programmes, the Commission has cooperated closely with other bodies such as the OECD, the Council of Europe, and the United Nations. It has ensured that Community legislation has kept pace with wider international thinking and, at the same time, that its voice is heard when agreements are being negotiated in international fora.

The large number of international conventions and agreements, to which the Community has in the process become a party, cover a wide range of environmental topics. They include the Convention on the Protection of the Rhine, the Geneva Convention on Air Pollution and the Barcelona Convention on the Protection of the Mediterranean. In the case of the Mediterranean, in particular, the Community is, for geopolitical, economic and cultural reasons, directly affected by its pollution and development problems. It is, therefore, imperative that it should make a more effective and specific contribution to the sound management of resources in that region, and to solving such problems by introducing any steps which it feels to be appropriate under the action plan for the Mediterranean to which it is a signatory. The Commission will also ensure that practical use is made of the opportunities for action concerning the environment which spring from Euro-Arab dialogue.

Given the transboundary nature of many environmental problems, and the impact that certain national measures can have on the economies of and trading relations with other countries, the environment has already become a regular topic for discussion at bilateral level between the Commission and a large number of countries.

A major test of the Community's credibility will come in the form of its efforts to deal with the issue of trade and transport in toxic wastes. At the present time serious differences divide nations and regions. There are those who call for a complete ban in the trade of toxic waste between developed and developing countries. There are others who say that while such trade should be kept to a minimum, it is important that, in partnership, third world countries should be able to mount the capacity to deal with toxic waste, particularly as they will require the wherewithal to face this challenge as their own industrial development advances. I believe that a compromise between these conflicting points of view is possible and needs to be negotiated. What is equally clear is that such a compromise would require the inclusion of the doctrine of prior informed choice, so that any third world country would have the ability to reject, as a right, any trade in toxic substances which it did not wish to accept. To play its part in achieving a worthwhile agreement, the twelve member states of the Community must themselves be rather more concerned with the sensitivities and understandable reluctance of third world countries to be exposed to risk, particularly in the light of the appallingly irresponsible incidents which have occurred, highlighted by the cases of 'The Karin B' and 'The Atlantic Conveyor'. If there is to be any trade at all it must be stringently controlled.

In this matter it is essential that the twelve Member States of the Community should adopt an altogether more responsible attitude than that manifested when they refused to countenance a Community draft directive which was designed to give Third World countries the opportunity to say no to the importation of chemicals which were severely restricted or prohibited altogether in the Community. This was done in order to protect the health of our 350 million people. On that occasion the Member States chose to protect the interests of their own chemical companies at the expense of the health and lives of people in Third World countries who are exposed to grave peril because of the abuse and excessive use of fertilisers and pesticides.

The ordinary citizen has a role to play at the very heart of our global challenge. At first sight the magnitude of the problem is such that one would question whether individual actions can have any effect. The fact is that there is no hope for dealing with that large array of challenges described in this book unless ordinary people in developed countries and in the Third World alike spring to the defence of the environment. It is ordinary people who will make governments give a high priority to the major international initiatives which are capable of answering these problems. But more people must become aware of the issues, more people must lobby governments for changes in policy, and more people must support practical, often small-scale initiatives. This, indeed, was the message that the European Commission sought to convey when promoting the European Year of the Environment. That year officially came to an end in March 1988 but, in reality, its purposes will never cease.

A few months ago, I came across a contemporary commentary on the scriptures. I would like to commend it to you.

> We waste what we have, our food, our fuel, our wealth, our gifts. And then we watch in surprise the destruction of our world. What we do not explode or gouge out of the earth, we pollute, and what we do not pollute, we kill. We do not see, or wish to see, the damage we do, and later we regret.[1]

The question that is posed to all of us is, are we simply going to regret or are we going to see?

14 The United Nations System and Sustainable Development

Javier Perez de Cuellar

The world is approaching a significant turning-point: the end of a period of confrontation between super-powers and the beginning of a new era in which peace will be the result of mutual confidence instead of the balance of nuclear terror. Rather than focussing on issues of short-term survival in the face of possible nuclear extinction, we can now dispose ourselves towards more long-term development and the fulfilment of human needs. The new challenge facing the international community is the eradication of poverty and the need for all countries to have the opportunity to develop in a world of shrinking resources. People everywhere must have the possibility to lead rewarding lives and enjoy the fruits of their labours. To achieve this, patterns of development must be sustainable, not only economically and financially, but also environmentally.

The Charter of the United Nations enshrines a fundamental truth in expressing the relationship between actions that maintain peace and security, and those which reaffirm faith in the dignity and worth of the human person, which establish conditions of justice, and which promote social progress and better standards of life of greater freedom. The United Nations was entrusted with the responsibility of serving as a centre for harmonising the actions of nations in the attainment of these common ends.

In a sense, the current discussion on sustainable development is a reaffirmation of this fundamental truth. The international community has come to recognise the imminent threat to human survival posed by environmental degradation. But equally important, it has recognised that an adequate response to this threat must include complementary action to promote sustainable patterns of growth, especially in developing countries. Needless to say, the wanton destructiveness of war is anathema to this effort, as is the violation of human rights.

The threats posed to the environment are of concern to all countries, North and South, East and West. But is it clear that no country or group of countries can respond alone to this challenge. The requirement is for action at national, regional and international levels. In order for them to succeed, these actions must be part of a comprehensive, coherent and co-ordinated strategy.

The United Nations system has an important role to play in the formulation and implementation of a strategy for sustainable development. As a universal forum, the United Nations encompasses many of the hopes and aspirations of mankind. Moreover, the various entities of the United Nations system possess an impressive array of talent and experience relevant to all aspects of the promotion of sustainable development.

The preservation and protection of the environment has long been a major activity of the United Nations. As far back as 1968, the General Assembly expressed its concern about the continuing and accelerating impairment of the quality of the human environment and 'the consequent effects on the condition of man, his physical, mental and social well-being, his dignity and his enjoyment of basic human rights, in developing as well as developed countries' (General Assembly resolution 2398 (XXIII)). By 1970, major environmental initiatives such as UNESCO's Man and the Biosphere (MAB) programme were underway in United Nations agencies. The 1972 United Nations Conference on the Human Environment was convened in Stockholm to consider environmental protection and improvement. The Conference's major concerns, which included human impact on the environment, the relationship between quality of the environment and quality of life and the connection between environment and development, were:[1]

(i) improvement of human settlements and health:
- creation of decent habitats for rapidly growing populations
- protection of human health;

(ii) development and use of fresh water, land and energy resources:
- availability of water for common use
- maintaining of soil fertility
- management of forest and mineral resources
- reconciling energy demands and environmental concerns;

(iii) harmonising development goals and social and cultural values with environmental quality objectives:

- relationship between development goals and environment
- relationship of social and cultural values and environment;

(iv) protection of living resources and of the oceans and avoidance of inadvertent climate modification:

- protection of terrestrial ecosystems, wildlife, genetic resources, and fisheries
- protection of the oceans
- man's impact on climate.

I believe that the Declaration adopted by the Conference, which includes principles designed 'to inspire and guide the peoples of the world in the preservation and enhancement of the human environment', represents the first global strategy in this area.

Three institutions were created by the General Assembly to carry out the important enterprise initiated at the Stockholm Conference: the Governing Council, a legislative environmental body composed of representatives of fifty-eight governments; an Environment Fund to support wholly or in part the costs of environmental initiatives undertaken within the United Nations system; and an environment secretariat. This last became known as the United Nations Environment Programme (UNEP) and was established to serve as a focal point for environmental coordination within the United Nations and a catalyst for environmental action.

From the beginning, it was recognised that the various components of the system had vital contributions to make within their respective areas of competence. The main tool for the co-ordination of these efforts is the System-Wide Medium-Term Environment Programme (SWMTEP), designed by UNEP in consultation with its partners, and approved by the executive heads of the agencies and organisations of the system. The SWMTEP allows the United Nations agencies to take environmental factors into account throughout the development of their activities. It also provides a coherent, overall framework by which the United Nations' resources are coordinated and brought to bear on urgent environmental problems. In 1982, the UNEP Governing Council approved the SWMTEP for the period 1984–89. Environmental considerations were also factored into the medium-term plans of individual agencies, including UNEP, and into the overall United Nations medium-term plan. A second system-wide plan, including thirteen programmes and sub-programmes and covering the period 1990–5, was subsequently approved in March, 1988.

In 1982, the UNEP Governing Council met at Nairobi in a session of a special character to commemorate the tenth anniversary of the Stockholm Conference. The session *inter alia* called for the establishment of a special commission to propose long-term environmental strategies for achieving sustainable development up to the year 2000 and beyond. This was endorsed by the General Assembly.

We were indeed fortunate in the quality and commitment of the individuals who came to serve on this special commission – the World Commission on Environment and Development – under the chairmanship of the former Prime Minister of Norway. Its Report, *Our Common Future*, has served to provide a new sense of purpose and direction to our deliberations as to how we can broaden the development options for the present generations while keeping choices available for future generations. The guiding principles of policy actions put forward in the report are not necessarily new but are given coherence by it. The principles are: to revive growth, change the quality of growth, conserve and enhance the resource base, ensure a sustainable level of population, reorient technology and manage risk, integrate environment and economics in decision-making, reform international economic relations, and strengthen international co-operation.

At its 43rd session, in 1987, the General Assembly welcomed the report of the World Commission. It agreed to the principles set out above and asserted the need for them to be set against a background for the preservation of peace and the remedying of poverty.

At the same session, the General Assembly adopted by consensus the *Environmental Perspective to the Year 2000 and Beyond*, a broad framework to guide national action and international co-operation for environmentally sound development, which drew heavily on the report of the World Commission and shared its broad conclusions, including its emphasis on sustainable development (General Assembly resolution 186 (XLII)). This resolution is of particular importance since, on the one hand, it reflects the collective wisdom of both experts and practitioners, including the organisations of the United Nations system, and, on the other, the political commitment of all Member States. It thus provides an important basis for translating the concept of sustainable development into reality.

The General Assembly has called for the convening of a world conference on environment and development in Brazil in 1992. This conference will serve to renew national and international commitment to a strategy for sustainable development. More importantly, I am convinced that the conference will provide a basis for concrete action.

The United Nations system, long active across the whole spectrum of social and development activity, has a clear mandate and responsibility to play a leading role in the achievement of sustainable development. United Nations agencies provide policy, scientific and expert advice as well as technical assistance and financial support to governments. They make major contributions in strengthening and training personnel in developing countries and in developing global scientific research and monitoring capabilities.

In 1988, I asked all components of the UN system to report on the measures they had taken to implement the recommendations of the World Commission on Environment and Development and the *Environmental Perspective*. Several broad conclusions can be drawn.

First, it is evident that the activities of the system in support of sustainable development originated long before the current debate began. The relationship between environment and development was a major theme of the Stockholm Conference and was emphasised in the Conference's Action Plan for the Human Environment. It has remained a central component of UNEP thought since the first session of the General Council, in 1973. It was also emphasised in the *World Conservation Strategy: Living Resource Conservation for Sustainable Development*, a programme commissioned by UNEP and prepared by the International Union for Conservation of Nature and Natural Resources (IUCN) to help advance the achievement of sustainable development through the maintenance of essential ecological processes and life-support systems, the preservation of genetic diversity, and the ensurance of the sustainable utilisation of species and ecosystems. The *Strategy* was published in 1980 after being endorsed by UNEP, the Food and Agriculture Organisation of the United Nations (FAO) and the United Nations Educational, Scientific and Cultural Organisation (UNESCO), as well as IUCN and the World Wildlife Fund (WWF) (now the World Wide Fund for Nature).

Second, the entities in the system have responded and developed new programmes and activities in support of sustainable development.[2] For example, the International Bank for Reconstruction and Development (World Bank) has integrated environmental considerations into the mainstream of the Bank's operational and policy work through a more systematic consideration of environmental issues on a country-by-country basis, through environmental monitoring of all aspects and at all stages of Bank-financed projects. It has also increased funding of environmental projects, and increased attention to environmental matters throughout the Bank's research, training and information activities. UNEP and the UN Economic Commission for Africa (ECA) brought together African environment, economy, planning and education ministers in June 1989, to discuss the achievement of sustainable development in Africa, and similar regional conferences will be organised by the UN Economic Commission for Europe (ECE), the UN Economic and Social Commission for Asia and the Pacific (ESCAP), and the UN Economic Commission for Latin America and the Caribbean (ECLAC). Perhaps most importantly, sustainable development underlies the second System-Wide Medium Term Environment Programme (1990–5). The following are its priorities:[3]

(a) Careful assessment, integrated planning and management of terrestrial and marine ecosystems for sustainable productivity, maintenance of biological diversity and attainment of food security, with priority given to controlling soil degradation, including desertification, improving the national management of tropical forests, extending and improving the coverage of co-operative programmes for the protection of oceans, the protection and management of coastal areas, the rehabilitation of depleted fish stocks and the management of river and lake basins;
(b) Improved assessment of conditions of the atmosphere with emphasis on reaching a better understanding of the dynamics of the ozone layer and assessing the likely environmental and socio-economic impacts of climatic change, and development of international legal instruments, as appropriate;
(c) Continuing assessment of the occurrence and effects of toxic and other dangerous substances in the environment, including pesticides and herbicides, and of hazardous, including nuclear, technologies and wastes; dissemination of relevant information, development of national and international institutions and management procedures to limit harmful effects, establish environmentally sound practices and prevent and control accidents, and development of international legal instruments, as appropriate;
(d) Development of less wasteful, more energy-efficient and ecologically acceptable technologies for industrial and agricultural development and promotion of their use in both developed and developing countries, with particular emphasis on meeting domestic energy needs in developing countries;
(e) Development and implementation of strategies, *inter alia* in the form of pilot projects at the national and regional level, for incorporating environmental considerations into the planning and

management of human settlements, including nomadic settlements;
(f) Dissemination of information to the public and support of environmental education and training to build up understanding of the interactions between people and the environment and the significance of environmental management, and, by changing the world-wide community, devolving and internalising environmental responsibility;
(g) Development and strengthening of the ability of the United Nations system and, more importantly, of Governments to take environmental factors into account from the beginning of and throughout the development process, by improving economic analysis techniques to better reflect environmental benefits and by improving institutions and procedures.
(h) Development of inventories of environmental constraints for incorporation into development strategies; and
(i) Development and inclusion in the evaluation processes of both the United Nations system and Governments of appropriate criteria and indicators which adequately reflect the overall long-term social and environmental impact of economic activity and administrative and legal measures.

Third, the activities of the system in this area have benefited greatly from increased public interest in the issues. The question of sustainable development now figures prominently on the agenda of the various inter-governmental bodies and was a major theme of debate at the 43rd (1988) and 44th (1989) General Assemblies. These new priorities have also been reflected in UN budgets, and medium-term plans.

Finally, the diversity of interests reflected in the mandates of the various components of the system permits a broad range of assistance to governments in their attempts to be environmentally sound and to develop sustainably. This is essential, given the multifaceted nature of the challenge, and will require a high degree of co-operation, coordination and cogency in our activities. But I am encouraged from my consultations with senior colleagues that this commitment to common effort exists and will guide our actions.

For the foreseeable future, the struggle for sustainable development will be a priority item on the international agenda and thus a central feature in the activities of the United Nations system. Significant progress has been achieved in some areas. For example, the successful conclusion of the Montreal Protocol for global protection of the ozone layer and the Basel agreement to regulate and reduce international disposal of hazardous wastes provide a basis for further agreements on international measures to protect and improve the environmental base for sustainable development. On another front, progress achieved in resolving some regional disputes and the general improvement in East-West relations offers promise of greater international co-operation on a wide range of issues.

However, I remain deeply concerned that in the areas of economic and social development, the current situation and future prospects remain bleak. Unless decisive and deliberate action is taken by the international community to deal with these problems, in particular to promote growth and development of the developing countries, our efforts in other areas may well be in vain.

The realisation of sustainable development will not be an easy task. As has been emphasised, it will require mobilising the energies of all peoples, non-governmental organisations, and governments – in fact the whole international community. It will require action to ensure that all countries, and particularly the developing countries, participate fully in the deliberations on a strategy, that their diverse interests are reflected in the conclusions, and that they have the means to fully implement agreed action programmes. It will require the effective integration of local, national, regional and global activities. But, perhaps most importantly, it will require a change in societal attitudes about the way we live, consume and produce.

The challenge is indeed vast. I am greatly encouraged, however, that there is a growing awareness of the gravity of the problems and the need for concerted efforts to overcome them. The United Nations system will continue to assist the international community in reducing the impact of pollution and environmental degradation, rehabilitating ecosystems that have already suffered and, most important, promoting sustainable development so as to bequeath to future generations a secure planet where every human being has the opportunity to fulfil his or her full potential.

15 Diplomacy and Sustainable Development

Sir Crispin Tickell

The idea of sustainable development won international currency with the publication of the report of the World Commission on Environment and Development in 1987. For all its ambiguities the idea was of obvious value. No-one could deny that development – one of the cant words of the previous twenty-five years – had to be sustained if it was to have meaning; yet the introduction of the word 'sustainable' hinted clearly enough that much previous development had not been sustainable, and that some development, through abuse of the environment, could have done as much harm as good.

Thus the vocabulary of debate began to catch up with the underlying realities, as it usually does in the end. Once recognised, those realities have come into sharper international focus. Few wanted to believe the Club of Rome in 1970. The Stockholm Conference on the Human Environment in 1972 helped to bring out some of the central issues, and led to the creation of the United Nations Environment Programme (UNEP). But for many countries environmental hazards looked peripheral, or at best something about which only the industrial rich could afford to worry. The Global 2000 report commissioned by President Carter in 1977 and published in 1980 was virtually ignored. Yet evidence of the worsening global problems – from spreading deserts to pollution of land and water to changes in atmospheric chemistry – continued to accumulate, and unease about the various courses on which the world and its governments seemed to be set became widespread.

One of the results was the creation by the United Nations General Assembly in 1983 of the World Commission on Environment and Development which emphasised the idea of sustainable development. I was explained that by this was meant: a political system that secured effective citizen participation in decision-making; an economic system that could generate surpluses and technical knowledge on a self-reliant and sustained basis; a social system that provided solution to the tensions arising from disharmonious development; a production system that respected the obligation to preserve the ecological base for development; a technological system that could search continuously for new solutions; an international system that fostered sustainable patterns of trade and finance; and an administrative system that was flexible and had the capacity for self-correction.

Over the years many such Commissions set up to explore a wide range of issues had made little impact. But there was a feeling, borne out by subsequent events, that the impact of this one would be greater. Although the Commission set out a formidable range of major problems, it also offered a message of hope. The Report therefore generated more interest than some of its predecessors which had seemed to suggest only gloom and despair. Its title *Our Common Future* reflected a consensus by its twenty-two Commissioners who had listened as well as spoken. Their message came not only from government leaders, representatives of voluntary organisations, educational institutions and the scientific community, but also from less grand and ordinary people, who in the first as well as the last resort must be the true agents of change.

The report went to the United Nations General Assembly. The Assembly is more of a debating society than a parliament. It operates on a one-country one-vote basis, and its resolutions have no legal force. In the past it has had a bad reputation for rhetoric. But if agreement or loose consensus can be reached, it can act as a moral focus of world opinion. In the case of the Brundtland Commission Report, it was clear from the outset that most countries, including the 128 of the 159 Members which describe themselves as 'developing countries' and belong to the 'Group of 77' (the G77), wanted to arrive at some sort of consensus. Most important they recognised not only a common interest but a multiplicity of national interests which pushed them towards doing so.

Prolonged debate and discussion led eventually to the adoption of Resolution 42/187 of December 1987. In this resolution the Assembly welcomed the Commission's Report and gave a general if qualified endorsement to the idea of sustainable development. It agreed that there should be an equitable sharing of the environmental costs and benefits of economic development, and it called somewhat vaguely upon governments and the international apparatus involved in the United

Nations system to carry out programmes to give effect to many of the Commission's recommendations. In the best United Nations tradition, much of the message had to be read between the lines. As is so often the case, the Resolution was as significant for what it did not say as for what it did: in particular industrial countries successfully resisted attempts by some members of the G77 to include an explicit rejection of 'environmental conditionality' (or new conditions attached to aid loans or grants from donor countries or the multilateral financial institutions). Instead the emphasis was laid on the need for policies based on mutual partnership and common endeavour to meet a global threat which affected everyone.

The passing of the Resolution marked a significant step in the approach of the United Nations towards environmental problems. Until 1987 the subject had been considered every two years. Resolution 42/87 included a paragraph to include in the agenda of the following – 1988 – Session of the General Assembly a new sub-item entitled 'A Long-Term Strategy for Sustainable and Environmentally Sound Development'. This made sure that the environment became a regular annual item. At the time of writing (November 1989), the Assembly has a range of environmental drafts before it, including one to prepare the way for a Conference on the Environment and Development in 1992.

The atmosphere in which these debates have taken place was markedly different from that of the recent past. The last half of the 1980s has seen three main trends, amid many positive and a few negative developments at the United Nations. First, the Soviet Union has made fundamental changes in its foreign policy, including a new attitude towards the United Nations, as expressed most vividly in President Gorbachev's speech to the General Assembly in December 1988. Cooperation of the kind envisaged in the Charter is not only possible but has effectively begun. Second, such cooperation has extended beyond East/West relations. The North/South confrontations which characterised the process of much of the 1960s and 1970s have been widely recognised as profitless and damaging. Third, there is growing awareness that global problems require global solutions and that the United Nations family of institutions provides the only framework for dealing with them.

Nowhere are these trends more evident than in the field of the environment. The North/South confrontations of the 1970s drowned the United Nations agenda with such initiatives as the New World Economic Order, the Charter of Economic Rights and Duties of States, and so on. The environment, then treated as a poor and doubtful relation, was scarcely admitted to the agenda. Now it is at the top. This does not mean that agreement on individual issues has become easy. Nor that fundamental differences do not continue to divide industrial and other countries. But the mood has become one of cooperation over tackling shared problems.

In looking to the future we must bear in mind that the United Nations can only reflect the realities of power, wealth, resources and geography. Different countries have markedly different capacities for change. It is no accident that the poorer countries are poor. They have fewer resources, and many of their inhabitants live in desert or tropical zones of high vulnerability. The complex ecosystems which sustain life on the edges of the desert or in the rain forests are extremely fragile, and once destroyed cannot be replaced. Many of their soils are poor, and cannot, without massive use of fertiliser (which brings its own problems), increase the supply of food. Those who have oil often do not have water. Those who have water see irrigation schemes silt and salt up. Rapid change in the customs of those who live in these environments, and the application of methods used successfully elsewhere, can all too often lead to economic and social disruption.

Population growth is perhaps the most serious problem of all. Current rates of increase in human numbers, with accompanying domesticated plants and animals, put at risk the very idea of sustainable development. As a species we are now consuming some 40 per cent of net primary production on land. None has done the like in biological history. Although agricultural productivity has more than doubled since 1950, use of fossil fuels has increased around seven times, thereby changing the chemistry of the atmosphere. For many countries keeping in step with an expanding population already makes economic progress almost impossible. Often encumbered by debt, and lacking the means to make best use of their limited resources, they become dependent on others for their survival. At present sub-Saharan Africa imports a growing proportion (over 10 per cent) of the cereal it consumes. In such circumstances protection of the environment, however essential, becomes extremely difficult. Survival until next week is more important than the survival of the next generation. If catastrophe is to be avoided, major international cooperation is required.

Part of the problem lies in the management of existing frameworks, for we have more assets, and more experience in using them, than is often realised. There is already a rich variety of institutions, some specialised, some more general in character, which cover environmental and development problems. There is no need to give a full list. But specialised bodies range from the international financial institutions (the World Bank, the International Monetary Fund and the regional development banks) to the United Nations Environment Programme and the United Nations Development Programme. More generally, there is the General Assembly and the Economic and Social Council. The performance of these bodies has varied greatly over the years, and there has been lack of coordination. Efforts are now being made to put things right: in some cases more resources are needed (for example for UNEP), and in others poor organisation and overlapping functions have been a major handicap (for example with the Economic and Social Council). At present, attention is being focussed on how best to organise and prepare for the World Conference on Environment and Development in 1992. These preparations should be a catalyst for creative change.

UNEP, based in Nairobi, will obviously have a major role at the Conference. It has done its best within its existing resources, but nearly everyone agrees that it requires strengthening. The World Meteorological Organisation has done an admirable job over the years, and, with UNEP, has promoted such bodies as the World Climate Programme, which is due to hold its second conference in 1990. Reporting to it is the Intergovernmental Panel on Climate Change which should lay out the scientific basis for future international work and decisions on this subject. The Intergovernmental Panel has already proved itself to be so useful that its life should be prolonged after the submission of its report to the World Climate Programme.

But these are essentially technical bodies. The problems of environment and development go so deep and wide, and touch so many of the ways in which modern society is organised, that the central organs of the United Nations must also be directly involved. The General Assembly is already seized by environmental and development problems, and will continue to discuss them at its annual meeting every autumn. The job of the Security Council, perhaps the most effective of the United Nations institutions, is to maintain international peace and security. There can be no question that if deterioration of the environment should cause, or seems likely to cause, friction or conflict between states, the Security Council would be required to take the necessary action. It requires little imagination to foresee increasing problems between states over, for instance, food, fresh water, and refugees. How far the mandate of the Council might be interpreted in this respect is a matter of discussion and argument. The International Court may also become increasingly involved. Litigation on environmental problems is already a growth industry.

There have been suggestions for new institutions or adaptations of old ones: for example some have argued for a new supranational authority. Others have suggested that the almost moribund UN Trusteeship Council, hitherto responsible for winding up colonialism, should be converted into a Trusteeship Council for the Planet with wide environmental responsibilities. But this would require a major change in the Charter, with all that would involve, and most states are at present unwilling to contemplate it. Another idea is to look back to a precedent set in 1946 when the world was seeking to come to terms with the phenomenon of nuclear energy. An Intergovernmental Commission on the subject was then set up by the General Assembly, reporting on certain points to the Security Council. Unfortunately it did not survive the strains of the cold war, but some of the ideas examined within it eventually found their way into the International Atomic Energy Agency.

In the debate leading up to the World Conference on Environment and Development in 1992, these and other ideas will certainly be considered. It is too early to say whether decisions on them can be taken in 1992. In the meantime it is perhaps more useful to look into the issues at the heart of the debate. All agree that global problems need global treatment; but some precedents for dealing with global problems are better than others. Few would like to follow the almost interminable and in the end unsatisfactory course of the Law of the Sea Conference. This was a major disappointment and a failure that is still being felt. Better is the precedent set by the way the international community dealt with the menace of chlorofluorocarbons and the damage they cause to the ozone layer. Under the guidance of the United Nations Environment Programme, work took place which led to the Vienna Convention of 1985, the Montreal Protocol of 1987, and current efforts to eliminate production of these chemicals altogether. Fortified by the outcome of the Montreal Protocol, the British government recently proposed to the United Nations that there should be a convention on climate change to set out guidelines, or a code of 'good climate behaviour'. Into this convention could be fitted special undertakings or protocols governing use of the main greenhouse gases. These protocols would need proper supervision and monitoring. Acceptance of obligations under them would have to be universal to prevent countries stealing a march on each other or pursuing beggar-my-neighbour policies.

There could also be a convention on the conservation of species. One of the most frightening phenomena of our time has been the disappearance of species at a rate which can only be compared to some of the great disasters in the earth's history millions or hundreds of millions of years ago. The diversity of life, of which we are a tiny part, is critical not only for the future of our own species but also to that of life itself.

This new awareness of the global character of the earth's problems does not make for easy diplomacy. All major global problems are linked in one way or another. But if we string together debt, disarmament, development, growth and the environment, and insist that they all be discussed on every agenda, little progress can be made on any of them. In preparing for the World Conference in 1992, we have to distinguish and work on the main problems, while taking account of the linkages they have with others. Unlike other species, we have changed, and are still changing, the face of the earth to meet our specific short-term needs. Those needs are limitless. But we live on a small planet, and the limitations on us are evident wherever we look. We now have to accept and exercise our responsibilities.

Part Six: Conclusions

16 Interpreting the Signals

Sir Arthur Norman

All of Planet Earth's inhabitants, to one degree or another, rely on instinctive skills in the competitive struggle for survival. In the natural world signals are continuously transmitted and received in a myriad of ways, and if they are disregarded or falsely interpreted the penalty, more often than not, is the forfeiture of life itself.

Mankind in the early stages of human development shared many of these sharp instincts, even if they never merited comparison with those of lower-order species with built-in navigation systems, or the ability to delay pregnancy or to self-abort when famine and overpopulation threatened. Modern man still possesses instinctive abilities but to these has been added a highly developed intelligence which arms him with even more powerful means of survival and progress. Information – the efficient collection, storage, retrieval, analysis and dissemination of facts – is now the main weapon used to chart his course towards security and prosperity.

The Brundtland Report, like others which preceded it, categorises the several danger signals being so clearly emitted from the natural world, and readers may be justified in questioning whether mankind's ability – or perhaps the will – to receive and interpret unwelcome messages has declined to a truly dangerous degree. The Report does not discuss that possibility, maintaining the positive stance from which it prescribes remedies and an action programme which the Chairman of the World Commission has described as an 'Agenda for Global Change'. How and for what reasons has it come about that mankind now requires such urgent prompting to change course?

To a human population of one billion, halfway through the nineteenth century, the inventions and discoveries of the latter part of the eighteenth century had brought the fruits of what Engels named as the 'Industrial Revolution'. The process of unlocking the secrets of nature and the development of a whole new range of scientific knowledge and practical skills were transforming Britain – and other countries to follow – from predominantly agricultural to predominantly industrial nations. The whole process of wealth creation had been transformed, and the consequences were many and great; they included an acceleration in the growth of population, new and higher standards of choice and consumption and, of course, new forms of industrial pollution, some of which have left indelible marks on the face of this country. There is evidence to suggest that the successes and benefits of industrialisation had given many decision-takers an enhanced feeling of mastery of the natural world, from which they began to distance themselves in a variety of ways.

Care for the natural environment, although more necessary than ever due to the events of that time, was entrusted to a small minority of dedicated naturalists and was accorded a low priority in human affairs. Nevertheless, in 1872 the first of the world's National Parks – Yellowstone – was established, and at the turn of the century a conference on the protection of birds was held in Paris. A few years later President Theodore Roosevelt convened a World Conference at the Hague, to discuss the best means of protecting global energy resources and to draw up an inventory of natural resources vital to the economy of the world. The International Office for the Protection of Nature was set up in Brussels in 1929. But these early signs of the beginning of a more general concern for the environment were not much in evidence when at the end of the Second World War the United Nations System saw the light of day. The UN's many new responsibilities and departments did not include special arrangements for care of the environment.

It was an English biologist, Sir Julian Huxley, who, as the first Director-General of UNESCO, noticed the omission, and who in 1948 assisted at the birth of what is now the International Union for Conservation of Nature and Natural Resources (IUCN). That organisation became the world's scientific point of reference in matters concerning the natural world, but scarcity of funds and the need to publicise and act upon IUCN's recommendations led to further action, which in 1961 established WWF as the first major non-government organisation with an international remit in the environmental arena. By 1972 evidence of rapidly accelerating damage to the natural environment led to the Stockholm Conference on the Human Environment, which is generally recognised as the event which laid the foundation of the strong and widespread environmental 'movement' of today. Among the results of that international event were the setting-up of the United Nations Environment Programme (UNEP) and the establishment of a new non-government organisation (NGO) named as the International Institute for Environment and Development (IIED), which as its name implies was designed to make the case for integrating environmental and economic considerations in the planning and implementation of development. By 1980 IUCN and WWF combined forces to produce and publish the World Conservation Strategy (WCS). Although this document was distributed to all international institutions and to all national governments and was strongly publicised, its main recommendation – the compilation of National Conservation Strategies – was taken up by relatively few nations, most of which were developing countries with slender means available for converting words into action.

Thus it was a long, tortuous and bumpy road which led in 1983 to the setting-up of the World Commission on Environment and Development, and to the publication of the Brundtland Report two years ago. Why can we be optimistic that this document and its clear and purposeful recommendations will not suffer the fate of its predecessors, which include the North/South report of the Brandt Commission? Will words be turned into action throughout the globe, and what form should that action take?

One answer to the first of these two questions is that within the last few years the force of public opinion on environmental issues, particularly in the industrialised democracies of the northern hemisphere, has become manifestly stronger, better founded on scientific evidence of real hazards, and more insistent in its demands for action than at any time in the past. There is now wider recognition that this tide of public opinion is still becoming more powerful and it has already demonstrated, in some countries, the ability to influence voting patterns, to halt a great new industry in its tracks, to change agricultural policies and practices and to strangle the markets for products which are known to be contributors to major environmental risks.

No political party and few decision-takers in the business world can believe that they are insulated from the effects of this new pressure for changes which, if implemented, would change much else. Political and business leaders have always relied upon their ability to read, interpret and act upon the signals coming from their market places; few of them would deny that their most costly errors and lost opportunities can be attributed either to poor research and observation, or to false deductions leading to mistakes of concept or timing, which have handed competitive advantage to others. In practice political and business managements have begun to make adjustments to policy and practice – optimism is justified.

In shaping action programmes of the kind indicated in the Bruntland Report, it becomes quickly evident that the agenda is genuinely global and that it requires the involvement of many diverse actors. The major forces of the natural world do not respect national or even regional boundaries, so the requirement is for international, regional, national and local programmes which produce action from governments, from public and private enterprises and from individual citizens in every part of the globe. The magnitude of such a task rules out any ideas of rapid changes brought about by a series of isolated 'quick fixes'. Progress on such a wide front will be gradual and continuous in nature, and considerable importance will be attached to efficient and regular monitoring, to information flows and to public education programmes. Heavy investment in new or modified methods and higher operating costs in the short-term will need to be accommodated progressively and smoothly in the uninterrupted march towards economic and social objectives which will become ever more urgently necessary as the human population continues to increase and to demand higher standards.

THE ROLE OF GOVERNMENTS

Governments in industrialised democracies are best fitted to assume the leading role in the process of change and redirection, for they have the economic and institutional strength to achieve results in their own economies which will inevitably have worldwide impact. Their programmes should include the following essential elements.

(1) Clear policies which recognise that a global agenda for change requires attitudes and conditions which promote determined collaboration at international and national levels, in addition to effective national programmes in which public authorities – central and local – as well as private enterprises and individual citizens are encouraged to participate.

(2) The adoption of well-publicised pro-active programmes for integrating environmental aims with economic and social aims in development planning at all levels. This implies a change, in many countries, from a reactive stance in which measures to safeguard the environment emerge only after damage has occured, and as

concessions made with reluctance and without clear conviction that they are a necessary contribution to the social and economic welfare of present and future generations.

(3) More substantial investment in environmental research analysis and monitoring. This element is a necessary safeguard against the adoption of wasteful and ineffective measures, and is a prime requirement for the accurate evaluation of progress towards clearly defined objectives.

(4) The overhaul and up-dating of the legislation and enforcement procedures which govern the control of pollution and waste products and which promote environmentally-sensitive development at all levels of human activity. This process should include the use of practical incentives and the use of the fiscal system to speed up the modification or elimination of activities which are identified as major contributors to environmental damage.

(5) The active promotion of trade policies, aid policies, debt servicing policies and other measures which assist the developing nations of the South to escape from the poverty trap which is a cause of environmental degradation on a massive scale.

These recommendations are not a recipe for interventionist government. Government is required to provide leadership and to create the conditions in which vigorous economic activity can progressively reduce the demands and burdens placed on the natural ecosystems which are necessary for human survival and progress. However, they do require a rapid change from some current policies and actions, which, in a competitive world economy, have been adding to the now widely publicised dangers to the health and safety of humanity.

THE ROLE OF INDUSTRY AND COMMERCE

The checklist of actions required from industry and commerce should include the following:

(1) The compilation, publication and progressive implementation of environmental, health and safety objectives which are designed to produce products, services and working conditions which protect the interests of customers, employees, investors and local communities, and which enhance the reputation of the enterprise.

(2) Action to ensure that the structure of a company's decision-taking framework fully incorporates environmental considerations at the earliest stages of all new developments, and that existing activities are continuously monitored and, where necessary, modified in the light of new knowledge and changing environmental circumstances.
(3) Progress reports to the company's constituents on environmental protection measures, which are given the same importance and prominence as employment policies and actions.
(4) The incorporation of environment protection in training programmes at all levels in the enterprise.
(5) The development of waste prevention, recycling and disposal policies and programmes which reduce wastage, improve costs and conform strictly to legal and regulatory requirements. Such action to include monitoring of the activities of sub-contractors.
(6) The adoption of packaging distribution and transport policies which minimise pollution, noise and inconvenience to communities affected by the company's operations. This action requires the maintenance of vehicles to a high standard and insistence on similar standards for sub-contractors.
(7) Action to ensure that the design and development of new technologies, processes and products take into consideration the strong probability that environmental laws and regulations will become progressively more demanding as time passes. Laboratory and field testing procedures which ignore this assumption may lead to costly modification and even to product withdrawal.
(8) Recognition, within the business planning and marketing processes, of the competitive advantage which can be secured from environmentally-sensitive products services and processes. Environmental improvements offer market opportunities of many kinds in the world's marketplace.
(9) The development and constant up-dating of project appraisal techniques which attribute proper values to environmental goods and effects and which aim to minimise detrimental effects on the resource base.

Most of these recommendations and many other measures of environmental consequence have already been adopted by the world's leading companies, but there are many enterprises which have yet to recognise the fact that all human activity has some effect on the environment. These recommendations are based on the belief that environmental problems and their solution result from many individual investment decisions, and on many individual actions taken in the course of implementation. The costs of investment which embraces environmental safeguards will usually be higher than the costs of those which do not, but the price of later modification, withdrawal or closure can be prohibitive and even destructive.

THE ROLE OF ORDINARY CITIZENS

In the industrialised democracies in the northern hemisphere the demands of ordinary citizens for environmental improvements and safety from health risks constitute the driving force which endows the Brundtland Report with the power to secure action from governments and from business communities. Citizens play the roles of taxpayer, customer, producer and investor and, most importantly, they are the possessors of votes in the political process.

To an increasing degree members of the general public are acting as custodians of their own environment, as the rapid growth of 'nimbyism' – the not-in-my-backyard syndrome – indicates. Self-interest is the most powerful of all motivations, but there are also indications that citizens are unwilling to abandon or reduce their demands for a safer environment in the face of ill-judged threats that they will have to foot the bill for environmental improvement. There are some who argue that this phenomenon will disappear when the bills are presented, but the evidence of recent opinion polls does not support this contention. The major test of the strength of the general public's attitude to environmental issues will be whether the day-by-day actions of individual members of society contribute or fail to contribute to the achievement of improvements in environmental quality. There are notable differences between one country and another in attitudes and practices relating to waste disposal, cleanliness, energy conservation, driving disciplines, noise abatement, home maintenance and many other areas in which individual decisions act for or against environmental quality. Family traditions, education and positive response to public campaigns organised by central and local governments are major influences on behavioural trends. The ordinary citizen will be called upon to do far more than express opinions on environmental issues.

Non-governmental charitable organisations (NGOs) represent the collective strength of ordinary citizens, who are the source of their membership and practical support. Their influence in many countries is considerable, and is on the increase in promoting environmental protection and improvement. Shrill publicity and exaggerated alarms are sometimes the tools of a minority of NGOs, but there is evidence of much powerful influence exerted by organisations which base their advocacy on careful scientific research and analysis. Where these conditions exist new partnerships are replacing old confrontations; partnerships are now being formed between governments and NGOs and between NGOs and industry and commerce, resulting in numerous practical projects and in effective remedies for identified ills. This is the way forward for those who understand the fact that environmental quality and safety is determined by the behaviour of all the actors on this stage.

The Brundtland Report is addressed to a world audience and it identifies the plight of the inhabitants of the poorest communities of many countries as a major source of environmental degradation. In pointing to the connection between population growth, poverty and pollution the Report is asking a crucially important question. If, today, with a world population of 5 billion of which almost 1 billion have only the barest means of survival, we are deeply concerned with the growing hazards of environmental deterioration, how can we expect to manage and overcome such problems when the human population may have increased to 8 or 9 billion in a mere fifty years' time?

That question demands an answer which can only be a positive one if there is now urgent and concerted action of a nature and scale which has only been seen in the face of major threats to peace and security. It provokes questions about the structure and strength and motivation of our international, regional and national institutions which are required to produce action programmes for worldwide implementation. The conquest of poverty, diminution of population growth, improvement of education and training in the use of productive and benign new technologies constitute a very large task, calling for the deployment of large resources. Experience has shown us that solutions cannot be imposed from on high – they will only be found through the involvement and motivation of all the people, everywhere – even in the poorest communities.

Here in the United Kingdom where the process of industrialisation had its origins, there is a treasure-house of accumulated experience in addressing opportunities and problems of every kind. This experience has provided the country with great seats of learning and with a wide variety of knowledge and skills, many of which will be invaluable in the search for practical solutions to current environmental problems. The UK is an island nation in which, because of its small size, careful and skilful land use is a basic requirement for the sustenance of a relatively large population with high standards. And, as an island nation, understanding and knowledge of the seas and oceans of the world has been a fundamental necessity. Few nations are better armed to play a leading role in taking up the challenge of Brundtland.

Getting one's own house in order, however, is a prerequisite of being in a position to influence the behaviour of others on the global stage. Some explanation is necessary of the fact that, despite a high level of industrial and agricultural activity in a small land area, environmental laws and regulations and the enforcement of these are widely considered to be more lenient and tolerant than those of many other industrialised nations.

The answer may be found in the geographical position of the country which has enabled UK governments and UK-based industries to take full advantage of the natural 'sink' which is provided by the seas and fast-running currents which surround this island kingdom. In a competitive world it would be strange if such an advantage had not been exploited and jealously protected in regional negotiations. This has led to the growth of severe criticism from our European neighbours and to great reluctance to contemplate changes amongst those in this country who might lose precious advantages. Similar, if not the same, influences have been at work in the disposal of airborne wastes which have become unwelcome additions to the environmental problems of our northern European neighbours.

The signals now emanating from the natural world combine with complaints and pressures from neighbours to present the UK with a greater and more costly programme of change than some of our competitors who have learned to live with more restrictive environmental regimes. Pressures for a slow rate of change and even for prolonged delays have been and will continue to be strong. However, as the recent North Sea negotiations have demonstrated, the first steps in an action programme are being taken. Fine judgment and considerable skill will be required to determine a speed of change which keeps cost burdens at supportable levels without incurring the charge that the UK is not behaving as a good neighbour to its allies and major trading partners.

In considering alternative courses of action the economic penalties of enforced change are always easier to calculate than the value, social as well as economic, of such change. The discipline of environmental economics is in the comparatively early stages of development, but there are recent initiatives in the UK which give encouragement for the future.

IIED and University College London are at work on a programme in Economics for Sustainable Development, and the UK Centre for Economic and Environmental Development is also engaged in the task of making the economic case for environmentally sensitive development.

Action programmes attuned to the recommendations of the Brundtland Report are in hand in specific areas. There is now a growing consensus among political and business leaders that economic development programmes must be geared to the necessity to achieve sharply lower pollution levels which can be accommodated within limits set by carefully monitored natural ecosystems. To ensure that progress is maintained and spread throughout the world a system of regular monitoring and appraisal needs to be established, and the results must be made known to the generators of the tide of public opinion which is now the driving force in a battle in which the people are leading their leaders.

In the past whole civilisations have faltered and collapsed because of failures to read, interpret and act upon the early signs of decay. The clear signals now reaching both leaders and the general public from the natural world are green – a colour which bids forward movement and action – not tomorrow or the next day, but continuously from now onwards.

17 Sustainable Development: Meeting the Growth Imperative for the 21st Century

Jim MacNeill

We live in a period of such rapid change that the past and the future are hardly on speaking terms. In the preceding chapters, distinguished leaders from around the world have documented a sudden acceleration of interrelated events on several fronts simultaneously – the economic, the ecological and the political. This is forcing profound changes in the relationships between peoples, nations and governments, and changes in the way we view and think about the management of the planet as a whole.

The world is on history's most rapid growth track, and the need for more sustainable forms of development has become an imperative. The key trends have become familiar. Since 1900, the world's population has multiplied more than three times. Its economy has expanded twenty times. The consumption of fossil fuels has grown by a factor of thirty, and industrial production has increased by a factor of fifty. Most of that growth, about four-fifths of it, has occurred in just the last thirty-nine years since 1950.

The pace of future growth could be even swifter, driven in part by a further doubling of world population within the lifetime of today's twenty-year-olds. Governments could act to stabilise population at a lower level, but their efforts to date clearly do not measure up to the challenge. If human numbers do double again within the next fifty years, a further five-to ten-fold increase in economic activity would be required to enable them to meet their basic needs and minimal aspirations – aspirations are as important as needs.

The gains in human welfare during this century have been enormous. If we continue to avoid world-scale conflict, the potential for future gains is even more awesome. Biotechnology, just one new branch of engineering, could itself change the world as we know it. Information technology already has. We now have potentially unlimited access to information and the advent of global communications makes it possible for people to begin to exercise responsibility for every part of the planet.

Glasnost, perestroika and the tidal shift in East-West relations have also opened the doors of opportunity. For more than forty years, world affairs have been dominated by the contest between East and West. The goal of each was to contain the expansion of the other. Recent changes have not only deprived both sides of their main enemy, they have begun to release the energies of the superpowers and their allies, and have made it possible for them to cooperate meaningfully on the critical issues of global change and human survival.

The processes of development that have produced such enormous gains in human welfare are also provoking major unintended changes in the earth's environment. Millions of people in dozens of countries have suddenly become aware that much of what God created, man is now destroying – not only Earth's basic life-supporting capital of forests, species and soils, but also its fresh waters and oceans, and even the ozone shield which protects all life from the sun's more deadly rays. Now we threaten ourselves with a rapid rise in global temperatures and sea-levels – greater, perhaps, in the next forty to sixty years than in the 10,000 years since the last ice age. The dismal trends have been well documented in the previous chapters and I will not repeat them.

The growth of the last forty years has been concentrated in the North. With only 25 – soon 20 – per cent of the world's population, industrialised countries consume about 80 per cent of the world's goods. That leaves more than three-quarters of the world's population with less than one quarter of its wealth. And the imbalance is getting worse, leading to increasing tensions with the South and causing growing numbers of people to become poor and vulnerable. Pervasive poverty at a time when we have the means and experience to eradicate it is the greatest single failure of any civilised society. It is also both a major cause and a major effect of environmental degradation and economic decline.

The industrialised countries are responsible for about 80 per cent of the world's pollution, and probably the same proportion of the rapid depletion of the Earth's ecological capital, even though much of that depletion occurs in the South. Today's financial flows and trading patterns result in a massive transfer of the environmental costs of Gross World Product from the richer industrialised countries to the poorer resource-based economies of the Third World. A study conducted for the Commission estimated these costs at about $14 billion a year — more than one third of the total amount of development assistance flowing annually in the other direction. Even this $14 billion is a low estimate because it only includes costs related to environmental pollution, not those related to resource depletion.[1]

The cumulative debt of developing countries has now reached roughly one trillion dollars; the interest payments amount to $60 billion a year. The traditional net flow of capital from developed to developing countries was reversed in 1982; more than $43 billion annually is now transferred in the other direction.[2] And that is only what the World Bank counts.

These bleak trends have shattered the simple faith of people in the permanent order of nature. Recent titles on the best-seller lists, and the opinion and editorial pages of the Western press abound in titles with terminal metaphors, metaphors like the 'end of nature', the 'end of death', the 'end of forever'. Everyone who thinks about it recognises that there is a very high probability that most of the world's children will not know the richness and variety of Earth that poets have celebrated down through the centuries. The world's economic and political institutions are seriously out of step with the workings of nature.

THE SUSTAINABILITY QUESTION

By 1982, a decade after the landmark Stockholm Conference on the Human Environment, it had become clear that environmental destruction, at a pace and scale never experienced, was undermining prospects for economic development, and threatening the very survival of Earth's inhabitants.[3] A year later, in the autumn of 1983, the United Nations General Assembly decided to call for the establishment of a special, independent commission, later named the World Commission on Environment and Development, to undertake a global enquiry and bring back some practical recommendations for change.[4]

Secretary General Perez de Cuellar invited Dr Gro Harlem Brundtland, then leader of the Opposition and later Prime Minister of Norway, to be Chairman of the Commission, and Dr Mansour Khalid, one-time Foreign Minister of the Sudan, to be Vice-Chairman. Dr Brundtland called me in Paris, where I was Director of Environment for OECD, and asked me to become a member (*ex-officio*) and Secretary-General of the Commission and to direct and manage its work. While this chapter draws heavily on the Commission's report, *Our Common*

Future, it reflects my own interpretation, as well as events since 1987 and information that has become available in the past two years.[5]

The question before the Commission was clear, if only tacit. Is there any way to meet the needs and aspirations of the 5 billion people now living on Earth, without compromising the ability of tomorrow's 8 to 10 billion to meet theirs?

To answer that question, the Commission went through a broad process of analysis, learning and debate. We contracted papers, established panels, and invited world figures to meet with us. We also did something that no previous international commission has attempted: we organised open public hearings in every region of the world, from Jakarta to Moscow, Sao Paulo to Oslo, Harare to Ottawa. We met and took evidence from nearly a thousand experts, political leaders and concerned citizens in five continents. In the process, we learned first-hand of the heavy contradictions between the reality of environment and development, which are totally interlocked in the daily lives of people, industries and communities, and the artificial distinctions drawn between the two by academic, economic and political institutions.

During its three years of work the Commission returned constantly to the question of sustainability. If Earth is already crossing certain critical thresholds, how is it going to accommodate a further five- to ten-fold increase in economic activity over the next fifty years? A five- to ten-fold increase over fifty years sounds enormous, but given the magic of compound interest, it actually reflects annual rates of growth of only 3.2 and 4.7 per cent. No government, developed or developing, aspires to less than that and, in fact, in many developing countries, it is hardly enough to keep up with projected rates of population growth, let alone reduce levels of poverty.

But it translates into a colossal new burden on the ecosphere. Imagine what it means in terms of planetary investment in housing, transport, agriculture, industry and every other part of our economic infrastructure.

Take energy, for example. If nations were to employ current forms of energy development, energy supply would have to increase by a factor of five, just to bring developing countries, with their present populations, up to the level of consumption now prevailing in the industrialised world. Critical life-support systems would collapse long before reaching those levels.

The question of sustainability has been forced front and centre by the acceleration of events. The answers to it are not evident because the obstacles to sustainability are not technical. They are mainly social, institutional and political. Economic and ecological sustainability are still dealt with as two separate questions in all governments and international organisations, where they are the responsibility of separate agencies such as ministries of finance and departments of environment. But economic and ecological systems are in fact interlocked. Global warming is a form of feedback from the Earth's ecological system to the world's economic system. So is the ozone hole, or acid rain in Europe and eastern North America, or soil degradation in the prairies, deforestation and species loss in the Amazon, and many other phenomena.

SUSTAINABLE DEVELOPMENT

After three years of enquiry in all parts of the world, the Commission concluded that a transition to sustainable forms of development is possible. But it will require a fundamental re-orientation of certain dominant modes of decision-making, in government and industry, and on the part of individuals who buy, consume and dispose of the world's goods. It will also involve significant changes, some of them politically difficult, in many of the economic, fiscal, energy, agricultural, trade, security and foreign policies which guide national, corporate and private behaviour.[6] Can these changes be made – and in time? The answers to that question are by no means evident either.

Since no one can predict the future, no one can rule out progressive ecological collapse. After all, the Four Horsemen are at work in parts of Africa, Asia and Latin America, fed in part by growing population/environment pressures. Threats to the peace and security of nations from environmental breakdown are today greater than any foreseeable military threat from conventional arms. They are increasing at a frightening pace. Local, even regional conflicts based on environmental disruption, water and other resource scarcities could well become endemic in the world of the future.

It is easy and sometimes tempting to get absorbed in prophecies of doom, but it really does not accomplish anything. The Commission preferred instead to explore the possibility, as we saw it, of a 'new era of growth' – not the type of growth that dominates today, but sustainable development, based on forms and processes that do not undermine the integrity of the environment on which they depend.[7]

An essential condition for sustainable development is that a community's and a nation's basic stock of natural capital should not decrease over time. A constant or increasing stock of natural capital is needed not only to meet the needs of present generations, but also to ensure a minimum degree of fairness and equity with future generations.

Another essential condition for sustainable development concerns the nature of production. If growth rates of up to 3 or 4 per cent in the industrialised countries, and up to 5 or 6 per cent in developing countries are to be sustained, a significant and rapid reduction in the energy and raw material content of every unit of production will be necessary. At the same time, we must invest heavily not only to maintain, but also to increase our stocks of ecological capital, so that future dividends can be increased.

The maxim for sustainable development is not 'limits to growth'; it is 'the growth of limits'. Some limits are imposed by the impact of present technologies and social organisation. But many present limits can be expanded through changes in modes of decision-making, changes in some domestic and international policies, and through massive investments in human and resource capital.

There is one form of growth that we do have to limit, however, and that is the growth of our own species. A sustainable future is conditional upon a significant and rapid reduction in high rates of population growth. The issue, of course, is not simply one of numbers. A child born in a rich, industrialised country places a much greater burden on the planet, than one born in a poor country. The industrialised world found that various processes of development were the best means of population control. The same processes are at work in developing countries and some of them are beginning to take strong direct measures to increase social, cultural and economic motives for couples to have small families. But these efforts are not receiving the financial and, what is perhaps just as important, the political support they urgently need from developed countries.

INTEGRATING ENVIRONMENT IN ECONOMIC DECISION-MAKING

The language of sustainable development is the integration of environment in economic decision-making.[8] Our economic and ecological systems are now totally interlocked in the real world, but they remain almost totally divorced in our institutions. This is one of the greatest barriers to sustainable development. During the '60s and '70s, governments in over 100 countries, developed and developing alike, established special environmental protection and resource management agencies. But governments failed to make their powerful central economic, trade and sectoral agencies in any way responsible for the environmental implications of the policies they pursued, the revenues they raised and the expenditures they made.

The resulting balance of forces was and is grossly unequal. Environmental agencies must now be given more capacity and more power to cope with the effects of unsustainable development policies. What is much more important, however, is that the environment be brought into the centre of economic decision-making. To that end, governments will need to make their central economic, trade, and sectoral agencies directly responsible and accountable for formulating policies and budgets to encourage development that is sustainable. Sustainability should become the test of sound economic policy.

This will require a number of changes. The most important of these, certainly the most fundamental, is in the way we count economic activity.[9] True national income is 'sustainable' income. When a nation's stocks of man-made assets depreciate, they are written off against the value of production and new investments are made to build up and maintain the stock of capital. At the moment, this does not apply to a nation's stock of ecological capital.

National economic accounting systems are concerned mainly with the flow of economic activity, and changes in stocks of ecological capital are largely ignored. If we were to integrate economic and resource accounts, society would be able to determine whether a reported increase in the gross domestic product is real, or not. It may well reflect a corresponding decline in the nation's stocks of soils, forests, fisheries, waters, parks and historic places.

We also have to take another look at the market. Most leaders have finally discovered that the market is the most powerful instrument available for driving development. What many have not yet discovered is that it can drive development in two ways – sustainable and unsustainable. Whether it does one or the other is not a function of an 'invisible hand', but of man-made policy. The market's inability to deal with externalities is well known. It treats the resources of the atmosphere, the oceans and the other commons as 'free goods', and it 'externalises' or transfers to the broader community, the costs of air, water, and land pollution, and of resource depletion. We have not learned how to deal with that yet, at least not politically.

Moreover, 'normal' government interventions in the market often distort the market in ways that pre-ordain unsustainable development.

Tax and fiscal incentives, pricing and marketing policies, exchange-rate and trade protection policies all influence the environment and resource content of growth. Yet those responsible for setting such policies seldom consider their impact on the environment or on stocks of resource capital. When policy-makers do take these things into account, they often assume implicitly that the resources are inexhaustible, or that substitutes will be found before they become exhausted, or that the environment 'should' subsidise the market. The same is true of certain sectoral policies.

Energy is a good example. If future energy needs are to be met without adding further intolerable pressures on the world's climate, nations will have to make efficiency the cutting-edge of their energy policies. Productivity gains of 1 to 2 per cent a year are quite realisable and would buy the time needed to increase the use of renewable forms of energy and to develop some new and more benign forms of power. It would also improve a nation's macroeconomic efficiency and international competitivity. Good economics and sound ecology can be – and often are – mutually re-inforcing.[10]

The major obstacle to this is the existing framework of incentives for energy development and consumption. These incentives are all-pervasive; they are backed by enormous budgets; and they usually promote the very opposite of what is needed for a sustainable energy future. They underwrite coal, shales, oil and gas; they ignore the costs of polluting air, land and water; they favour inefficiency and waste; and they impose enormous burdens on already tight public budgets.[11]

Deforestation is another good example. Government policy cupboards are full of incentives to overcut the world's forests. Brazilian taxpayers underwrite the destruction of the Amazon to the tune of hundreds of millions a year in tax abatements for uneconomic enterprises. Indonesians do the same. So do Canadians. American taxpayers are subsidising the clearing of the Tongass, the last great rain forest in Alaska. If these economically perverse and ecologically-blind incentives remain in place, most of the world's remaining forests will probably not survive, with all that implies for food security, deserts, flooding, and global warming.[12]

Soil and water degradation is another example of unwittingly destructive economic incentives. Virtually the entire food cycle in North America, Western Europe and Japan attracts huge direct or indirect subsidies. These subsidies encourage farmers to occupy marginal lands and to clear forests and woodlands. They induce farmers to use pesticides and fertilisers in excess, and to waste underground and surface waters in irrigation. Canadian farmers alone lose well over $1 billion a year from reduced production due to erosion stemming from practices underwritten by the Canadian taxpayer.[13]

According to the Organisation for Economic Cooperation and Development (OECD) and other sources, the farm subsidy structure now costs Western governments in excess of $300 billion a year. What conservation programmes can compete with that? These subsidies send farmers far more powerful signals than do the small grants usually provided for soil and water conservation. The adverse effects of these subsidies extend beyond national borders. By generating vast surpluses at great economic and ecological cost, the subsidies create political pressures for still more subsidies: to increase exports, to donate food as *non*-emergency assistance to Third World countries, and to raise trade barriers against imported food products. All of these measures hurt agricultural productivity in developing countries.[14]

Public policies that actively, if unintentionally, encourage deforestation, desertification, destruction of habitat and species, and decline of air, soil and water quality must be reformed. These policies, and the enormous budgets they often command, are much more powerful than any conceivable measures to protect environments or to rehabilitate those already damaged.

If governments must rig the market – and no politician that I have met is likely to give up that proven path to power – they should do so in ecologically sensible ways. Agricultural subsidy structures, for example, could be changed in ways that not only maintain farm income, which is clearly essential for sustainable agriculture, but also encourage practices that enhance rather than deplete basic farm capital. North American models for such policies go back to the 1930s, when the Soil Conservation Service in the US and the Prairie Farm Rehabilitation Administration in Canada brought the Dust Bowl under control. The US Farm Security Act of 1985 provides a more recent example of the type of changes needed.

Energy incentives can also be provided in ways that encourage conservation and end-use efficiency, and that discourage fuels that lead to acid rain and global warming. To do so, countries will have to consider 'conservation pricing', that is, taxing energy during periods when the real price is low to encourage increases in efficiency. Stricter regulations should demand steady improvements in the efficiency of appliances and technologies, from electrical motors to air conditioners, in building design, automobiles and transportation systems. Institutional innovation will also be necessary to break utility supply monopolies and to re-organise the energy sector so that energy services can be sold on a competitive, least-cost basis.

A nation's annual budget establishes the framework of economic and fiscal incentives and disincentives within which corporate leaders, businessmen, farmers and consumers make decisions. It is perhaps the most important environmental policy statement that any government makes in any year, because in their aggregate these decisions serve to either enhance or degrade the nation's environment, and increase or reduce its stocks of ecological capital.

A budget that levies taxes on energy, resource use and pollution, matched by an equivalent reduction in labour, corporate and value-added taxes, could have a significant impact on consumption patterns and on the cost structure of industry, without adding to the overall tax burden on industry and society. Reform of tax systems along these lines seems essential to encourage a transition to sustainable development.[15]

With increased awareness, the politics of changing incentive systems should not be insurmountable. Some leaders of the most advanced industries have welcomed analyses linking economic incentives and environmental integrity. Providing their income is not jeopardised, farmers have everything to gain from incentive systems that encourage practices that maintain or enhance their soils, wood, water and other basic farm capital. For consumers, many such shifts in incentives would be neutral, and the impact on employment could even be positive.

REFORM OF INTERNATIONAL INSTITUTIONS

The Commission's report is, above all, a strong call for new and stronger forms of international cooperation. The massive changes occurring in the relationships between the world of nation states and the earth and its biosphere, have not been accompanied by corresponding changes in our international institutions.

Legal regimes, for example, have been rapidly outdistanced by the accelerating pace and scale of change. New norms for state and interstate behaviour are needed. At a minimum, the world community requires some basic norms for prior notification and consultation with neighbouring states when domestic developments have significant international consequences. States must accept an obligation to alert and inform their neighbours in the event of an accident likely to have a harmful impact on them – Basel, Chernobyl, Exxon Valdez – as well as an obligation to compensate them for any damage done. Similar norms and obligations must gradually evolve for the global commons and future generations.

Some progress has been made on specific and relatively narrow issues since the report came out in April, 1987. Examples include the 1987 Montreal Protocol on Ozone, the 1988 Protocol on NOx, and the 1989 Basel Convention on Hazardous Wastes. Overall, however, international cooperation to address global threats from the perspective of environmental protection remains highly inadequate.

International cooperation to address these threats from the perspective of sustainable development has hardly started. Most global environmental threats are fueled by so-called 'domestic' development policies. International trade and monetary policy can have an enormous impact on environmental degradation and resource depletion in other countries, but they have usually been treated as 'domestic' affairs.[16] Energy, agricultural, industrial and other incentive policies are discused and studied in fora like OECD, but they have seldom been made the subject of negotiations for mutually agreed change.

A serious attack on high rates of population growth, on the social and economic reforms needed to address poverty (such as education, endowment of increased power to women, protection of human rights and land reform), and on the reforms needed to address climate change, deforestation, species loss, soil erosion, and marine pollution will all require a lot more international attention to these 'domestic' policies of sovereign states. These policies and the interests they serve are no longer 'national' – they reach into the back yards of other states and global commons.

It will also require new ways of marshalling financial resources for sustainable development in the Third World. The Commonwealth Prime Ministers' meeting in Kuala Lumpur was the first intergovernmental meeting, to my knowledge, at which world leaders talked about this question using numbers that bear some relationship to the needs.[17] On an initiative of Prime Minister Gandhi, they considered raising a sum of the order of 15 billion dollars. Hopefully, they have added three zeros to future discussions of financing international action on environment and sustainable development.

Debt remains the most urgent problem facing developing countries, particularly those in Africa and Latin America. The debt problem must be resolved before these countries can be expected to turn their attention to the pressing agenda of poverty and interlocked economic and ecological decline. Although several plans for debt relief (most recently the US Brady Plan) have been advanced, they all share two dubious characteristics: the conventional measures attached as a condition for additional loans; and the absence of any reference to programmes to sustain, let alone build up, the environmental resource capital of developing countries.

There are several possible international sources of revenue that could be tapped to finance action in support of sustainable development. The use of the international commons could be taxed, for example, as could trade in certain commodities. A World Atmosphere Fund has been proposed that would be financed, in part, by a 'climate protection tax'. Revenues would come from a levy on the fossil fuel consumption of industrialised countries, and the proceeds would go to developing countries to help them limit and adapt to the consequences of global warming and sea-level rise. Others have proposed that the tax should be related to the carbon content of fuels. Most recently, the Norwegian Government proposed that, as a starting point, industrialised countries allocate 0.1 per cent of their GNP to such a fund. The recent Netherlands' budget included provision for an annual contribution of 250 million guilders to a 'global climate fund', and the Netherlands government has assessed the various options for financing and managing such a fund.

Military expenditures also represent an enormous pool of capital, human skills and resources, a large proportion of which could well be shifted to more productive purposes. That would require greater awareness of the growing scale of environmental threats to national and regional security, an awareness that some political leaders are beginning to voice. It would also require a new and broader concept of security, a concept that encompasses environmental as well as economic and political security. With a broader approach, nations would begin to find many instances in which their security could be improved more effectively through expenditures to protect, preserve and restore basic environmental capital assets than it could through expenditures for arms.[18]

A world in which power is being rapidly diffused by education and by the mobility of information, capital and people, requires multiple forms of international governance. The Commission put forward a large number of proposals for reforms in this direction. Others have since been added to the list. Some take an incremental approach, building on existing institutions; others propose more fundamental change, involving some pooling of national sovereignty. There are some signs that the time is ripe to reinforce this effort.

The Hague Declaration of March, 1989, for example, recommends a new international authority with responsibility to prevent further global warming.[19] It recognises that this can be done either by establishing a new institution or by extending an existing one. The most significant element in the Declaration is that it accepts that the institution should be able to take certain decisions by a means other than consensus. And it anticipates measures to enforce compliance, with appeals against such measures being placed before the International Court of Justice.

More than thirty nations have signed the Hague Declaration. It is not binding, but it should help the international community to break out of traditional mind-sets in judging other proposals for institutional change. Interestingly, the Paris Economic Summit in July 1989, while not going nearly as far as the Hague, did agree that 'new instruments may be contemplated' to protect the atmosphere.[20] Vague, perhaps, but unthinkable even a year ago, and pointing in the right direction.

Leadership is beginning to emerge, and a number of proposals have been advanced. Most of them focus on ways and means to strengthen the United Nations, although regional organisations like the OECD and ASEAN, and non-governmental institutions, would all have a vital role to play.

It has been proposed, for example, that the Security Council should periodically devote a special session to environmental threats to peace and security. The Soviet Union has hinted that it would support a new environment council, equal in authority to the Security Council, and perhaps without the right of veto. The Trusteeship Council is coming to the end of its existing mandate, and another proposal would transform it into a forum in which the nations of the world would exercise their 'trusteeship' for the integrity of the planet as a whole, including the global commons.

The idea of a new 'Earth Council', along the lines of one of these formulations, reflects the quality of imagination and the level of ambition that needs to be applied in developing new forms of governance to guide the planet through the coming turbulent decades. But I fear that it may be as much in advance of the current realities of international politics, as those realities are behind planetary trends.

Changes to strengthen the United Nations Environment Programme (UNEP) have also been proposed. It has been suggested that UNEP should be given the resources necessary to act on behalf of the Secretary General as the 'secretariat' for the Security Council, or a new Trusteeship Council. Some proposals have taken another tack. They would strengthen UNEP's mandate and resources in the area of environmental protection, but they would establish a new standing UN Commission on the Environment, or a Special Independent

Commission on Environment and Sustainable Development to service the 'Security' or 'Trusteeship' Councils through the Secretary General.

Countries must begin to treat the integrity of the environment and the sustainability of development as a foreign policy issue of paramount importance. Measures to reduce debt and to increase the net flow of resources to developing countries should be backed up with coherent policies on aid, on agricultural and other forms of trade, and on policies concerning the import or export of hazardous chemicals, wastes and technology. A 'foreign policy for environment and development' could help to induce greater coherence in these areas. It could also serve to improve overall effectiveness, coordination and cooperation concerning rapidly evolving developments in the management of the commons – the oceans, the atmosphere and outer space. This could be vital as we move into the next Ozone Conference, the World Climate Conference in late 1990, and the World Conference on Environment and Sustainable Development in 1992. Several leaders directly involved see the latter as the benchmark conference for the next decade.

POLITICS OF CHANGE

The world today is very different from the one that the Commissioners shared five years ago when we started our work. At our first meeting in Geneva in 1984, we speculated about the political climate that might prevail when our report was presented to the world community in three years' time. Would it be receptive to recommendations for significant changes in the way we manage the world's economy and environment? Or would it yawn and continue with 'business as usual'?

Our Common Future could not have appeared at a better time. During the past few years, a monumental change in public opinion has forced these issues to the top of political agendas in the United Nations, in Washington, London and other capitals around the globe, in some multilateral banks and in many of the board rooms of the Fortune 500. During the first full year after our report was presented to the General Assembly, more heads of state converted to environmentalism, within the new context of sustainable development, then converted to Christianity or Islam in any single century when conversion was assisted by the point of the sword.

At the Economic Summit in Paris, in July 1989, the entire Group of 7 endorsed the concept of sustainable development and vied with each other for environmental leadership.[21] The same thing happened three months later at the Commonwealth Prime Ministers' Conference in Kuala Lumpur.[22] Leaders in Eastern Europe have joined the green competition, as *glasnost* has uncovered layers of concern about decades of environmental neglect. So have leaders in the South, which is rapidly awakening to the fact that development cannot be maintained when basic ecological capital is drawn down relentlessly and without regard to its replenishment.

We may be living through one of those rare 'hinges of history', with unprecedented opportunities for policy and institutional innovation. The issues of environment and sustainable development are reshaping national and international affairs, and they may well become the overarching issues for the next century. Public opinion is far ahead of governments on these issues and global warming alone will ensure that it stays there. The politics of greening will continue to drive the greening of politics well into the twenty-first century.

Notes

1 ENDANGERED EARTH

1. Estimates of Worldwatch Institute, *State of the World, 1988* (London, 1988) p. 6 (based on FAO, *Tropical Forest Resources* (Rome, 1982) p. 5).
2. Centre for Science and Environment, *The State of India's Environment: 1984–85* (New Delhi, 1985) p. 80.
3. Report of World Commission on Environment and Development, *Our Common Future*, (Oxford, 1987) ch. 6. Also, E. Wolf, *On the Brink of Extinction: Conserving the Diversity of Life*, Worldwatch, Paper 78 (Washington, DC, 1988) pp. 101–7.
4. Examples from *State of the World*, op. cit., pp. 5–7.
5. These estimates from *World Demographic Estimates and Projections: 1950–2025* (New York, 1988) p. 246.
6. J. Jacobsen, 'Planning The Global Family' in *State of the World*, op.cit., p. 155. There is wider evidence pointing to unmet need in 'Fertility Behaviour in the Context of Development', evidence from *The World Fertility Survey* (United Nations Department of Economic and Social Affairs, New York, 1987).
7. These figures are from the UN Department of Economic and Social Affairs and the World Bank.
8. J. Jacobsen in *State of the World*, op. cit., p. 155.
9. Peter Usher (UNEP), *The Montreal Protocol on Substances that Deplete the Ozone Layer* (Ozone Depletion Conference, London, November 1988).
10. R. Warrick *et al.*, *The Greenhouse Effect, Climatic Change and Sea Level: An Overview* (University of East Anglia/Commonwealth Secretariat (mimeo), 1988) pp. 14–17.
11. Auberon Waugh, *The Independent Magazine*, 31 December 1988.
12. *The Observer*, 18 September 1988.
13. *The Economist*, 15 October 1989, p. 84.
14. R. Warrick *et al.*, op. cit.
15. *Case Studies of Pacific Atolls (J. Lewis) and Guyana (R. Camacho)* (Commonwealth Secretariat (mimeo) 1989).

16. Over and above those of displaced and otherwise affected (45 million in total) by flooding of the Ganges and Mahananda rivers.
17. *Our Common Future*, ch. 7; J. Goldenberg, T. Johansson, A. Reddy and R. Williams, *Towards an Energy Strategy for a Sustainable World* (World Resources Institution, Washington, DC 1987) pp. 29–38.

2 THREATENED ISLANDS, THREATENED EARTH; EARLY PROFESSIONAL SCIENCE AND THE HISTORICAL ORIGINS OF GLOBAL ENVIRONMENTAL CONCERNS

1. J.D. Hughes, 'Theophrastus as ecologist', *Environmental Review*, 4, (1985) pp. 296–307; see also C. Glacken, *Traces on the Rhodian shore; nature and culture in Western thought, from ancient times to the end of the eighteenth century* (Berkeley, Calif., 1967).
2. For example see D. Worster, 'The vulnerable earth; towards a planetary history', in D. Worster (ed.), *The ends of the earth; perspectives in modern environmental history* (Cambridge, 1988).
3. Author of *Man and Nature* (New York, 1864). This was one of the first texts to explore the history of environmental degradation and tc warn of the possible consequences were it to remain unchecked. See D. Lowenthal, *George Perkins Marsh; versatile Vermonter* (Cambridge, Mass., 1958).
4. See, for example, M. Williams, *The Americans and their forests; an historical geography* (Cambridge, 1989).
5. E.g. see K. Thomas, *Man and the natural world* (Oxford, 1983); T. O'Riordan, *Environmentalism* (London, 1976).
6. R.H. Grove, 'Conservation and colonial expansion; a study of the evolution of environmental attitudes and conservation policies on St. Helena, Mauritius and in India, 1660–1860' unpublished PhD thesis, University of Cambridge, 1988 (in press under same title, Cambridge University Press, 1990). See also paper under same title in *Past and Present*, 1990.
7. P.J. Marshall and G. Williams, *The great map of mankind; British perceptions of the world in the Age of Enlightenment* (London, 1982); B. Smith, *European vision and the South Pacific, 1768–1850; a study in the history of art and ideas* (2nd ed., Oxford, 1960).
8. Useful detailed discussions of the idealised new iconography of the tropics can be found in Leo Marx, *The machine in the Garden; technology and the pastoral ideal in America* (New York, 1964); and T. Bonyhady, *Images in opposition; Australian landscape painting 1801–1890* (Oxford, 1985).
9. S. Darian, *The Ganges in myth and history* (Honolulu, 1978).
10. S. Crowe and S. Haywood, *The gardens of Mughal India* (London: India Office Library, 1972).
11. J. Prest, *The Garden of Eden; the botanic garden and the recreation of Paradise* (New Haven, Conn., 1981).
12. W. Halbfass, *India and Europe; an essay in understanding* (New York, 1981) p. 21.
13. E. Bloch, *Das Prinzip Hoffnung* (Frankfurt, 1981).

14. E.g. see Lynn White, 'The historical roots of our ecological crisis', *Science*, 155 (1967) pp. 1202–7; and J. Opie, 'Renaissance origins of the environmental crisis', *Environmental Review*, 1 (1987) pp. 2–19.
15. In this connection it might be noted, for example, that rapid deforestation of the Ganges basin in pre-colonial northern India during the sixteenth century does not appear to have been impeded by indigenous religious factors.
16. Dante may himself have been informed by the Greek precedent of the Isles of the Hesperides, but would also have been aware of the attractions of the newly-colonised Canary Islands, at a time before they had been completely deforested.
17. M.C. Karstens, *The Old Company's Garden at the Cape and its Superintendents* (Cape Town, 1957).
18. See R.H. Grove, 'Conservation and colonial expansion', op. cit.
19. R. Bryans, *Madeira, pearl of Atlantic* (London, 1959).
20. P. Ligon, *A true and exact history of the island of Barbados* (London, 1673). Ligon noted, 'mines there are none in this island, not so much of as coals, for which reason we preserve our woods as much as we can'.
21. D. Watts, *The West Indies, patterns of development, culture and environmental change since 1492* (Cambridge, 1986).
22. e.g. Worster, *Nature's Economy* (Cambridge, 1977) pp. 29–55.
23. For details of this see J. Mackenzie, *The Empire of nature; hunting, conservation and British imperialism* (Manchester, 1989); and T. Pringle, *The Conservationists and the Killers* (Cape Town, 1983).
24. D. Washbrook, 'Law, state and agrarian society in colonial India', *Modern Asian Studies*, 15 (1981) pp. 649–721; the highly ambiguous attitude adopted by the early colonial government towards capital and the risks which its uncontrolled deployment entailed is a subject discussed very fully by Washbrook, although he is apparently unaware of the aptness of his arguments to the ecological dimension.
25. To date, only Lucille Brockway (in *Science and colonial expansion: the role of the British Royal Botanic Garden* (New York, 1979) and P. Mackay (in *In the wake of Cook; exploration, science and empire* (London, 1985)) have attempted to assess, on a global scale, the relationship between science, colonial expansion and commerce. Both writers attach exclusively utilitarian and/or exploitative and hegemonic motivations to the early development of science in the colonial (especially East Indian Company) context, and ignore the potential for contradictory or humanitarian motivations, of a kind certainly present among many early botanists in India, for example.
26. H.H. Spry, *Modern India* (London, 1837).
27. D.G. Crawford, *A history of the Indian medical service* (London, 1914).
28. S. Pasfield-Oliver, *The life of Philibert Commerson* (London, 1909). The first scientific 'Academie' was founded on Mauritius by Commerson in 1770.
29. R.H. Grove, 'Charles Darwin and the Falkland islands', *Polar Record*, 22 (1985) pp. 413–20.
30. R.H. Grove, 'Conservation and colonial expansion', op. cit.
31. Similarly, all three were early advocates of the abolition of slavery and highly critical of the corruption and absolutism of the *ancien régime*.

Poivre's collected works were published in 1797 as revolutionary tracts. After he left Mauritius, St Pierre went on to become the confidant of Rousseau and the first major French 'romantic' novelist.

32. R.H. Grove, 'Surgeons, forests and famine, the emergence of the conservation debate in India, 1788–1860', *Indian Economic and Social History Review* (1990).
33. Ibid. The Rajah of Nilumbur had as early as 1830 warned the Bombay Government of the dangers of uncontrolled deforestation, but his missives had been ignored. Later, however, they were taken up by Alexander Gibson in a report made to the Bombay government in 1840.
34. Draconian forest reserve regulations, which do not recognise the existence of traditional rights, have continued to be used by some post-colonial governments more interested in lining the pockets of corrupt ministers with timber interests than in protecting the environments of their own indigenous peoples. The scandalous persecution of the Penan and other forest peoples by the government of Malaysia is a case in point.
35. R.H. Grove, 'Colonial conservation, ecological hegemony and popular resistance; towards a global synthesis', in J. Mackenzie (ed.) *Imperialism and the natural world*, 1990.
36. R.H. Grove, 'Early themes in African conservation; the Cape in the nineteenth century', in D. Anderson and R.H. Grove (eds), *Conservation in Africa; people, policies and practice* (Cambridge, 1987) pp. 22–39.
37. J.F. Wilson, 'On the progressing desiccation on the Orange river in Southern Africa', *Proceedings of the Royal Geographical Society* (1865) pp. 106–9.
38. J. Spotswood Wilson, 'On the general and gradual desiccation of the earth and atmosphere', *Report of the Proceedings of the British Association for the Advancement of Science, Transactions* (1858) pp. 155–6.
39. A few years later, John Tyndall developed the concept of the 'atmospheric envelope' and the notion of 'greenhouse' retention of radiation heat by particular gases in the atmosphere. In doing so he built upon theories of heat transfer in the atmosphere first developed by Jean-Baptiste Fourier between 1807 and 1815; see J. Tyndall, *On radiation; the Rede lectures at the University of Cambridge, 16 May, 1865* (London, 1865).
40. This theory was elaborated on by Arrhenius in 1896 when he raised the possibility that increasing concentrations of carbon dioxide due to the burning of fossil fuel could lead to global warming. He calculated that doubling levels of carbon dioxide could raise average temperatures by 5 degrees C.
41. R.H. Grove, 'Conservation and colonial expansion', op. cit., p. 74 (St Helena Records, Council to EIC Court of Directors, April 9 1713, pp. 105–6).
42. For details, see multiple entries in the diaries of William John Burchell, unpublished MSS, The Hope Entomological Library, University of Oxford.
43. C. Waterton, *Wanderings in South America, the North-West of the United States, and the Antilles in the years 1812–1824* (2nd ed. London, 1880) esp. pp. 289–94. Here he writes of the forests of the Americas that 'Nature is fast losing her garb and putting on a new dress in these extensive regions... spare it . . . inhabitants . . . these noble sons of the forest beautifying your

landscapes beyond all description; when they are gone a century will not replace their loss, they cannot, they must not fall'.

44. C. Lyell, *Principles of Geology, being an attempt to explain the former changes of the earth's surface, by reference to changes now in operation*, 3 vols (London, 1834).
45. R.H. Grove, 'Scottish missionaries, evangelical discourses and the origins of conservation thinking in Southern Africa 1820–1900', *Journal of Southern African Studies*, 15 (1989) pp. 164–87.
46. E. Dieffenbach, *Travels in New Zealand* (London, 1843) on extinctions pp. 7–12, 50–2; on forest destruction pp. 257, 297–8; on the destructive impact of man, pp. 416–17.
47. H.E. Strickland, 'On the progress and present state of ornithology', pp. 213–15; H.E. Strickland and A.G. Melville, *The Dodo and its kindred* (London, 1848).
48. See M. Di Gregorio, 'Hugh Edwin Strickland on affinities and analogies', *Ideas and Production*, 7 (1987) pp. 35–50.
49. H. Cleghorn, *The forests and gardens of south India* (Edinburgh, 1861).

3 THE CHANGING CLIMATE AND PROBLEMS OF PREDICTION

1. Parts of this chapter were derived from: S.H. Schneider, *Global Warming: Are We Entering the Greenhouse Century?* (San Francisco, 1989) and S.H. Schneider, 'The Changing Climate', *Scientific American,* 261, no. 3 (September, 1989) pp. 70–9.

5 DEFORESTATION: A BOTANIST'S VIEW

1. Data from *Production Year Book, 1985*, vol. 39 (FAO, Rome, 1986).
2. J.-P. Lanly, 'Tropical forest resources', FAO Forestry Paper 30 (Rome, 1982).
3. J.-P. Malingreau and C.J. Tucker, 'Large-scale deforestation in the Southeastern Amazon Basin of Brazil', *Ambio*, 17 (1988) pp. 49–55.
4. J.-P. Malingreau and C.J. Tucker, 'The contribution of AVHRR data for measuring and understanding global processes: large-scale deforestation in the Amazon Basin', *International Geoscience and Remote Sensing Symposium* (Ann Arbor, Michigan, 1987), pp. 443–8.
5. E. Matthews, 'Global vegetation and land use', *Journal of Climate and Applied Meteorology*, 22 (1983) pp. 474–87.
6. J.-P. Lanly, op. cit.
7. G.T. Prance, W.A. Rodrigues and M.F. da Silva, 'Inventário florestal de um hectare de mata de terra Firme km 30 da Estrada Manaus-Itacoatiara', *Acta Amazonica*, 6 (1976) pp. 9–35.
8. A.H. Gentry and C. Dodson, 'Contribution of nontrees to species richness of a tropical rain forest', *Biotropica*, 19 (1987) pp. 149–56.
9. T.L. Erwin, 'Beetles and other insects of tropical forest canopies at Manaus, Brazil, sampled by insecticidal fogging', in S.G. Sutton, T.C. Whitman and A.C. Chadwick (eds) *Tropical Rain Forest: ecology and management* (Oxford, 1983), pp. 59–75.

10. A.H. Gentry, 'Endemism in tropical versus temperate plant communities' in M.E. Soulé (ed.), *Conservation Biology – The science of scarcity and diversity* (Sunderland, Massachusetts; Sinauer Associates, 1986) pp. 153–81.
11. S.A. Mori, B.M. Boom and G.T. Prance, 'Distribution patterns and conservation of eastern Brazilian coastal forest tree species', *Brittonia*, 33 (1981) pp. 233–45.
12. For birds, see J. Cracraft, 'Historical biogeography and patterns of differentiation with the South American avifauna: areas of endemism', in P.A. Buckley *et al.* (eds); *Neotropical ornithology* (Ornithological Monographs no. 36, American Ornithologists' Union, Washington, DC, 1985) pp. 49–84. For primates, see G.A.B. da Fonseca, 'The Vanishing Brazilian Atlantic Forest', *Biological Conservation* 34 (Elsevier Applied Science Publishers Ltd, Barking, 1985) pp. 17–34. For a general account of biological conservation see S.A. Mori, 'Eastern, extra-Amazonian Brazil' in D.G. Campbell and H.D. Hammond (eds), *Floristic Inventory of Tropical Countries* (New York: The New York Botanical Garden, 1989) pp. 427–54.
13. S.A. Mori, op. cit. See also, W. Dean 'Deforestation in southeastern Brazil' in R. Tucker and J.F. Richards (eds), *Global deforestations and the nineteenth century world economy* (Chapel Hill, North Carolina, 1983) pp. 50–67.
14. T.L. Lovejoy, 'A projection of species extinction' in *The Global 2000 Report*, vol. 2 (London, 1982), pp. 328–31. See also P.R. and A.H. Ehrlich, *Extinctions* (New York: Random House, 1982); N. Myers, 'Mass extinction: profound problem, splendid opportunity', *Oryx*, 22 (1988) pp. 205–10.
15. G.T. Prance, W. Balée, B.M. Boom and R.C. Carneiro, 'Quantitative ethnobotany and the case for conservation in Amazonia', *Conservation Biology*, 1 (1987) pp. 296–310.
16. B.M. Boom, 'Amazonian Indians and the forest environment', *Nature*, 314 (1985) p. 324.
17. *Ecological aspects of development in the humid tropics* (Washington, DC: National Academy of Sciences, 1982).
18. L.C. Brown, and E.C. Wolf, *Soil erosion: Quiet crisis in the World Economy*. Worldwatch Paper No. 60 (Washington, DC, 1984).
19. V.A. Villa Nova, E. Salati and E. Matusi, 'Estimativa da evapotranspiraçáo na Bacia Amazônica', *Acta Amazonica*, 6 (1976) pp. 215–28.
20. D. Janzen, *Guanacaste National Park: Ecological and cultural restoration* (San José, Costa Rica: UNEP, 1986).
21. R.J.A. Goodland and H.S. Irwin, *Amazon jungle: Green hell to red desert?* (New York: Elsevier, 1975). Also S.H. Davis, *Victims of a miracle: Development and the Indians of Brazil* (Cambridge: Cambridge University Press, 1977).

6 AGRICULTURAL POLLUTION: FROM COSTS AND CAUSES TO SUSTAINABLE PRACTICES

1. G.R. Conway and J.N. Pretty, *Fertiliser Risks in the Developing Countries* (International Institute for Environment and Development: London, 1988).
2. Ibid.
3. G.A. Norton and J.D. Mumford, 'Decision making in pest control', *Advanced Applied Biology*, 8 (1983) pp. 87–119.
4. S.K. De Datta, *Principles and Practices of Rice Production* (New York: John Wiley and Sons, 1981). Also, P.J. Greenwood, T.J. Cleaver, P.K. Turner, J. Hunt, K.B. Niendarf and S.M.H. Loquens, 'Comparisons of the effect of nitrogen fertiliser on the yield, nitrogen content and quality of 21 different vegetables and agricultural crops', *Journal of Agricultural Science*, 95 (1980) pp. 471–85.
5. National Research Council, *Regulating Pesticides in Food. The Delaney Paradox* (Washington, DC: National Academy Press, 1987). Also, D.G.R. Gilbert, 'Pesticide Control Policy and Safety Arrangements in Britain', unpublished PhD Thesis, University of London, 1987.
6. G.R. Conway and J.N. Pretty, op.cit. Also, M.L. Higgins, W.W. Barclay and J.N. Pretty, 'The Use and Management of Agricultural Chemicals in Asia and the Near East Region', paper prepared for the Environment and Natural Resources Strategy to USAID/ANE (Washington, DC, 1989).
7. M.L. Higgins, W.W. Barclay and J.N. Pretty op. cit.
8. R. Repetto, *Paying The Price: Pesticide Subsidies in Developing Countries* (Washington, DC: World Resources Institute, 1985). Also, M.L. Higgins, W.W. Barclay and J.N. Pretty op. cit.
9. P.E. Kenmore, F.O. Carino, C.A. Penez, V.A. Dyck, and A.P. Gutierrez 'Population regulation of the brown planthopper (*Nilaparvata Stal.*) within ricefields in the Philippines', *Journal of Plant Protection in the Tropics*, 1 (1984) pp. 19–37.
10. G.P. Georghiou, 'The magnitude of the resistance problem' in National Research Council, *Pesticide Resistance. Strategies and Tactics for Management* (Washington, DC: National Academy Press, 1986).
11. R. Barker, R.W. Herdt and B. Rose, *The Rice Economy of Asia. Resources for the Future* (Washington, DC, 1985).
12. J.D. Mumford, 'Farmers' attitude towards the control of aphids on sugar beet', *Proceedings of the 1977 British Crop Protection Conference — Pests and Diseases* (1977).
13. G.A. Norton and J.D. Mumford, 'Decision-making in pest control', *Advanced Applied Biology*, 8 (1983) pp. 87–119.
14. P.E. Kenmore, J.A. Litsinger, J.P. Bandong, A.C. Santiago and M.M. Salac, 'Philippine rice farmers and insecticides: thirty years of growing dependency and new options for change' in J. Tait and B. Napompeth (eds) *Management of Pests and Pesticides. Farmers' Perceptions and Practices* (Boulder and London: Westview Press, 1987).
15. I. Fagonee, 'Pertinent aspects of pesticide usage in Mauritius', *Insect Science and its Application*, 5 (1984) pp. 203–12.

16. Royal Commission on Environmental Pollution, Cmnd 7644, *Agriculture and Pollution*, 7th Report (London, 1979).
17. P.G. Fenemore and G.A. Norton, 'Problems of implementing improvements in pest control: a case study of apples in the UK', *Crop Protection,* 4 (1985) pp. 51–70.
18. A.C. Goldman, 'Agricultural pests and the farming system: a study of pest hazards and pest management by small-scale farmers in Kenya' in Tait and Napompeth, op. cit.
19. P. Slovic, 'Perception of risk', *Science*, 236 (1987) pp. 280–5. Also, Royal Commission on Environmental Pollution, Cmnd 9149, *Tackling Pollution — Experiences and Prospects. 10th Report* (London: HMSO, 1984).
20. P.W. Holden, *Pesticides and Groundwater Quality. Issues and Problems in Four States.* (Washington, DC: National Academy Press, 1986).
21. Her Majesty's Inspectorate of Pollution, *Annual Report* (Department of the Environment, London, 1989).
22. S. Fujisaka, *Participation by Farmers, Researchers and Extension Workers in Soil Conservation,* Sustainable Agriculture Programme Gatekeeper Series SA 16, International Institute for Environment and Development (London, 1989).

7 HALOCARBONS AND STRATOSPHERIC OZONE — A WARNING FROM ANTARCTICA

1. BrO
2. ClO
3. OH
4. CH_3Cl
5. CCl_4
6. $CFCl_3$
7. CF_2Cl_2
8. CH_3CCl_3
9. $C_2F_2Cl_3$
10. CHF_2Cl
11. CF_4
12. CH_3Br
13. $CBrClF_2$
14. $CBrF_3$
15. $C_2F_4Br_2$
16. CCl_4
17. $C_2F_2Cl_3$
18. CF_4
19. $CBrF_3$

8 CHANGES IN PERCEPTION

1. J. Lesourne, 'Reunion at Fontainebleau', IUCN *Bulletin*, 20 (1989) nos. 1–3, p. 4.

2. Sir James Fraser, *The Golden Bough* (London, 1890).
3. M.G. Royston, *Pollution Prevention Pays* (Oxford, New York, Toronto, Sydney, Paris, Frankfurt: Pergamon, 1979).
4. D.R. Harris, 'Swidden Systems and Settlement' in P.J. Ucko, R. Tringham and G.W. Dimbleby (eds), *Man, settlement and urbanism* (London: Duckworth, 1972).
5. J. Golson, 'No Room at the Top: Agricultural Intensification in the New Guinea Highlands' in J. Allen, J. Golson and R. Jones (eds), *Sunda and Sahel: Prehistoric Studies in Island Southeast Asia, Melanesia and Australia* (London: Academic Press, 1975).
6. J. Oates, 'Prehistoric Settlement Patterns in Mesopotamia', in P.J. Ucko, R. Tringham and G.W. Dimbleby (eds), op. cit.
7. T. Gabriel, 'Agricultural Human Resources' in R.W. Dutton (ed.), *The Scientific Results of the Royal Geographical Society's Oman Wahiba Sands Project, 1985–87, Journal of Oman Studies* Special Report 3 (Muscat: Diwan of Royal Court, Sultanate of Oman, 1988).
8. A.K. Biswas, M.A.H. Samaha, M.H. Amer and M. Abu-Zeid, *Water Management for Arid Lands in Developing Countries* (Oxford, New York, Toronto, Sydney, Paris, Frankfurt: Pergamon, 1980).
9. E. Ashby and M.E. Anderson, *The Politics of Clean Air* (Oxford: Clarendon Press, 1981).
10. R. Carson, *Silent Spring* (Greenwich, Conn: Fawcett Publications, 1962).
11. P. Ehrlich and A. Ehrlich, *Population, Resources, Environment. Issues in Human Ecology* (San Francisco: W.H. Freeman, 1970).
12. B. Commoner, *The Closing Circle* (New York: A Knopf, 1971).
13. P.B. Stone, 'The media and the teachers. A partnership for the environment' in *Insights: Learning for an Interdependent World* (Geneva, 1989).
14. M.W. Holdgate, M. Kassas and G.F. White (eds), *The World Environment, 1972–82* (Dublin: Tycooly International Publishing Ltd, 1982).
15. L.H. Sallada and B.G. Doyle (eds), *The Spirit of Versailles: The Business of Environmental Management* (Paris: ICC Publishing SA, 1986).
16. 'Better Environment Awards for Industry 1988', *RSA Journal* 138 (1989) no. 5397, pp. 533–43.
17. UNFPA 1987–89, *Populi: Journal of the United Nations Population Fund* (see particularly 14, no. 4, 1987).
18. 'Declaration of Fontainebleau', IUCN *Bulletin*, 20, nos. 1–3, p. 7.

9 RELIGION AND THE ENVIRONMENT

1. See above, pp. 12–14.
2. *Creation Eucharist*, Festival of Faith and the Environment, September 1989, pp. 10–13.
3. See above, p. 95.
4. David Gosling, 'The Morality of Nuclear Power', *Theology*, 81, no. 679 (January 1978) p. 27.
5. *Hymns Ancient and Modern, Revised* (London: William Clowes and Sons Ltd, 1950), p. 394.

6. For further details see David Gosling, 'Towards a Credible Ecumenical Theology of Nature', *Ecumenical Review*, 38, no. 3 (July 1986) p. 328.
7. Sally McFague, *Models of God* (London: SCM, 1987).
8. For further details see *Religion and Nature Interfaith Ceremony*, World Wildlife Fund, September 1986, and *The New Road*, 1 (Winter 1986/7), and subsequent issues.
9. For a concise summary of the conference see the report by Jeanne Knights to the St Francis Trust, October 1989; Ruth Page's contribution is summarised on p. 7.
10. *Creation Eucharist*, Festival of Faith and the Environment, September 1989, p. 5.
11. *Creation Harvest Liturgy* and *Creation and Harvest Service Book*, WWF and Winchester Cathedral, October 1987.
12. The speeches are reproduced in *Current Dialogue*, 10 (June 1986), p. 6.
13. *Celebrating One World: A Resource Book on Liturgy and Social Justice* (London: CAFOD and Thomas More Centre, 1987), and *Renewing the Earth, Study Guide for Groups* (London: CAFOD, 1987).
14. *Threads of Creation, A Resource Book of Words and Pictures*, edited by John Reardon (London: United Reformed Church, 1989); *Harvest Pack 1988: Making Peace with the Planet* (London: Methodist Church, Division of Social Responsibility, 1988).
15. *First Things First*, a Study Guide (London: BCC and Christian Aid, 1989).
16. *The Work of Our Hands* (London: Tear Fund, 1989).
17. *Death of a Forest* (Manila: Columban Mission, 1988) p. 2.
18. For further details, see David Gosling, 'Thailand's Bare-headed Doctors' *Modern Asian Studies*, 19, no. 4 (1985) pp. 761–96; same author, 'Visions of Salvation: A Thai Buddhist Experience of Ecumenism', forthcoming.
19. *Banking on the Poor* (London: Christian Aid, 1988); *For Richer for Poorer* (Oxford: Oxfam, 1986).
20. See p. 14 above.
21. Mother Mary Clare, *The Simplicity of Prayer* (London: SLG Press, Fairacres Publications, 1988).

11 OECD NATIONS AND SUSTAINABLE DEVELOPMENT

1. I wish to acknowlege the dedicated work and valiant assistance of Glen Okrainetz. He researched, double-checked, provided constructive criticism and kept me honest throughout this chapter.
2. The World Commission on Environment and Development, *Our Common Future* (Oxford: Oxford University Press, 1987).
3. The member nations of the OECD are: Austria, Australia, Belgium, Canada, Denmark, France, the Federal Republic of Germany, Finland, Greece, Iceland, Ireland, Italy, Japan, Luxembourg, the Netherlands, New Zealand, Norway, Portugal, Spain, Sweden, Switzerland, Turkey, the United Kingdom and the United States. The Socialist Federal Republic of Yugoslavia takes part in some of the work of the OECD.
4. Final Communiqué of the G-7 economic summit, June 1988. Toronto, Canada.

5. Michael Redclift, *Sustainable Development: Exploring the Contradictions*. (New York: Methuen, 1987).
6. Alex Davidson and Michael Dence (eds), *The Brundtland Challenge and the Cost of Inaction* (Ottawa: Institute for Research on Public Policy, 1988).
7. Susan Holtz, 'Environment-Economy Integration', *Canadian Business Review*, 16, no. 2 (Summer 1989).
8. William Ruckelshaus, 'Towards a Sustainable World', *Scientific American*, 261, no. 3 (September 1989).
9. The NGO community hosted conferences on wide-ranging topics including: 'The Legal Challenge of Sustainable Development' (Canadian Institute of Resources Law); 'Thinking Globally, Acting Locally in Burnaby' (Burnaby Citizens for Environmental Protection); 'Environment and the Economy: Partners for the Future', 'Renewables: a Clean Energy Solution' (Solar Energy Society of Canada); 'Misitu: the Environment of Africa's Sustainable Development' (Canada-Africa International Forestry Association).
10. Canadian Council of Resource and Environment Ministers, *Report of the National Task Force on Environment and Economy* (Winnipeg, 1987).
11. Text of the motion served by the Liberal Party of Canada and debated in the House of Commons on 15 May, 1987: 'That this House urge the Government to endorse, adopt and advocate, at home and abroad, the goals and recommendations of the report of the World Commission on Environment and Development (Brundtland Report), entitled *Our Common Future*; that this Report, dealing with global concerns, challenges and endeavours, be adopted as the central policy of the federal Government, its Agencies and Commissions; and that the Minister of the Environment advocates the spirit and the substance of this Report at the coming General Assembly of the United Nations on behalf of Canadians and of the global community, thus continuing Canada's tradition of giving leadership on environmental matters.'
12. Bill C-29, an Act to Establish the Department of Forestry, passed by the House of Commons on 1 November 1989. The Liberal Opposition amendment read: 'In this Act, . . . "sustainable development" means development that meets the needs of the present without compromising the ability of future generations to meet their own needs.' Section 6 of the Act obligates the Minister of Forests to '(d) have regard to the integrated management and sustainable development of Canada's forest resources'...
13. The Act to establish the Department of Industry, Science and Technology was passed by the House of Commons on 22 June 1989. The amendment would have included the pursuit of sustainable development in the new Department's mandate.
14. Sustainable Development Branch, Environment Canada, *Federal economic instruments and the achievement of environmental objectives* (Ottawa, 1989).
15. Ibid.
16. Sustainable Development Branch, Environment Canada, *Sustainable Development in ERDAs/EDAs, interim report* (Ottawa, 1988).
17. First Ministers' Conferences are meetings of the Prime Minister, ten

provincial premiers and two territorial government leaders. The agenda of the 9–10 November 1989 meeting held in Ottawa included environment and sustainable development.

18. Federal and provincial fisheries ministers agreed in July, 1989, to pursue a national sustainable fisheries policy and to develop a national position in preparation for the UN World Conference on Sustainable Development in 1992.
19. Agriculture Canada, *Growing Together: A Vision for Canada's Agri-food Industry* (Ottawa, 1989).
20. Canadian International Development Agency, *Environment and Development: The Policy of the Canadian International Development Agency* (Ottawa, 1987).
21. Canadian Environmental Advisory Council, *Canada and Sustainable Development: A Commentary on Our Common Future, the Report of the World Commission on Environment and Development and its implications for Canada* (Ottawa, 1987).
22. Science Council of Canada, *Environmental Peacekeepers: Science, Technology and Sustainable Development in Canada* (Ottawa, 1988).
23. *1989–90 Estimates: Part I: the Government Expenditure Plan* (Ottawa: Supply and Services, Canada 1990).
24. *Energy, Mines and Resources. 1989–90 Estimates: Part III:* (Ottawa: Supply and Services, Canada 1989).
25. Forestry agreements between the federal and provincial governments date back to the early 1980s. The intent is to provide federal money for reforestation and enhanced forest management techniques. As of November 1989, none of the six agreements that expired in March 1989 had been renewed.
26. OECD, *International Response to the Report of the WCED*. Prepared for the OECD seminar on the Report of the WCED, 17 November 1988 (Paris: OECD).
27. Meeting on the Protection of the Environment of the Conference on Security and Co-operation in Europe, 16 October to 3 November 1989, *Final Document: CSCE/sem. 36* (Sofia: 1989).
28. Royal Kingdom of the Netherlands, *National Environmental Policy Plan: To Choose or To Lose*, Second Chamber of the States General, Session 1988–9, 21 137, nos. 1–2 (The Hague, 1989).
29. Reports from a seventy-two-nation ministerial conference on global warming, held in the Netherlands in November 1989, indicate that the United States and Japan refused to agree to hold carbon dioxide emissions at 1988 levels by the year 2000, because they believe that further study is necessary before binding controls can be proposed.
30. Prime Minister Margaret Thatcher, address to the UN General Assembly, 8 November 1989: 'We should always remember that free markets are a means to an end. They would defeat their object if by their output they did more damage to the quality of life through pollution than the well-being they achieve by the production of goods and services.'

Recommended Reading

1. Organisation for Economic Co-operation and Development, *Environment and Economics* (Paris, 1984).
2. Neil Evernden, *The Natural Alien: Humankind, and the Environment* (Toronto: University of Toronto Press, 1985).
3. The Ministry of Environment, Norway, *Environment and Development: Programmes for Norway's Follow-up of the Report of the World Commission on Environment and Development*. Report to the Storting no. 46 (1988–9) (Oslo, 1989).
4. David Pearce, Anil Merkandya and Edward Barbier, *Blueprint for a Green Economy* (prepared for the UK Department of the Environment) (London, Earthscan Publications Ltd, 1989).
5. Robert Paehlke, *Environmentalism in the Future of Progressive Politics* (New Haven: Yale University Press, 1989).
6. Jim MacNeill, John Cox and David Runnalls, *CIDA and Sustainable Development* (Halifax, Canada: Institute for Research in Public Policy, 1989).

12 COMMON FUTURE – COMMON CHALLENGE: AID POLICY AND THE ENVIRONMENT

1. A.E. Housman, *The Collected Poems* (London: Jonathan Cape, 1939).
2. Oscar Wilde, *Works* (London: Collins, 1948).
3. Personal communication.
4. Personal communication.
5. C.P. Cavafy, *Collected Poems*, trans. E. Keeley and P. Sherrard, ed. G. Savidis (London: Chatto and Windus, 1978).

13 THE BRUNTLAND REPORT, ENVIRONMENTAL ADVANCE AND THE EUROPEAN COMMUNITY

1. 'The Day of Atonement', *The Reform Synagogue Prayerbook* (Oxford: Oxford University Press, 1985).

14 THE UNITED NATIONS SYSTEM AND SUSTAINABLE DEVELOPMENT

1. 'Action Plan for the Human Environment: Programme Development and Priorities: Report of the Executive Director (UNEP)', UNEP/GC/5 (2 April 1973) p. 3.
2. Document E/1989/L period 26, prepared by the executive director of UNEP on behalf of the UN Secretary-General and submitted to the General Assembly via the economic and social council. The report was issued in final form as document A/44/350-E/1989/99 and includes accounts from national governments and the UN Department of International Economic and Social Affairs (DIESA), UN Centre for

Human Settlement (HABITAT), UN Centre on Trans-National Corporations (UNCTC), UN Development Programme (UNDP), Economic and Social Commission for Western Asia (ESCWA), UNESCO, FAO, UN Industrial Development Organisation (UNIDO), International Atomic Energy Agency (IEAE), International Civil Aviation Organisation (ICAO), International Labour Organisation (ILO), International Maritime Organisation (IMO), UN Population Fund (UNFPA), World Bank, World Food Programme (WFP), World Health Organisation (WHO), and the World Tourism Organisation (WTO).

3. UNEP, *System-Wide Medium-Term Environment Programme 1990–1995*, UNEP/GCSS. I/7/Add. 1 (May 1988), pp 12–13. Item (i) has been paraphrased.

17 SUSTAINABLE DEVELOPMENT: MEETING THE GROWTH IMPERATIVE FOR THE 21st CENTURY

1. I. Walter and J.H. Loudon, 'Environmental Costs and the Patterns of North-South Trade', World Commission on Environment and Development, 1986.
2. World Bank, *World Debt Tables, External Debt of Developing Countries, 1988–89* (Washington, DC: The World Bank, 1989).
3. Martin W. Holdgate, Mohammed Kassas, Gilbert F. White, (eds) *The World Environment 1972–1982*, (Dublin: Tycooly International Publishing Ltd, 1982).
4. United Nations General Assembly, Resolution 38/161, 1983.
5. The World Commission on Environment and Development, *Our Common Future* (Oxford: Oxford University Press, 1987). See also Jim MacNeill, 'Strategies for Sustainable Economic Development', *Scientific American*, 261, no. 3 (September, 1989).
6. Ibid.
7. Jim MacNeill, 'Sustainable Development: What is it?' Statement to Joint Economic Committee, Congress of the United States, 13 June 1989.
8. Jim MacNeill, 'Strategies', op. cit.
9. David Pearce, Anil Merkandya, and Edward Barbier, *Blueprint for a Green Economy* (prepared for the UK Department of the Environment) (London: Earthscan Publications Ltd., 1989).
10. Jose Goldenberg, Thomas B. Johannson, Amulya K. N. Reddy, and Robert H. Williams, *Energy for Development* (Washington, DC: World Resources Institute, 1985).
11. M. Kosmo, *Money to Burn? The High Cost of Energy Subsidies* (Washington, DC: World Resources Institute, 1987).
12. R. Repetto, *The Forests for the Trees? Government Policies and the Misuse of Forest Resources* (Washington, DC: World Resources Institute, May 1988).
13. *Soil and Risk: Canada's Eroding Future*, A Report on Soil Conservation to the Senate of Canada (Ottawa, 1984).
14. For a full discussion of ecologically perverse agricultural policies and their reform see: *Food 2000, Global Policies for Sustainable Agriculture*, the

Report of the Advisory Panel on Food Security, Agriculture, Forestry and Environment to The World Commission on Environment and Development (London: Zed Books Ltd., 1987). Also, The World Commission on Environment and Development, *Our Common Future* (Oxford: Oxford University Press, 1987), Chapter 5; and the World Bank, *World Development Report, 1986, Part II, Trade and Pricing Policies in World Agriculture* (Oxford: Oxford University Press, 1986).

15. Professor Dr Ernst U. von Weizsacker, 'Global Warming and Environmental Taxes', International Conference on Atmosphere, Climate and Man, Torino, Italy, 16–18 January 1989.
16. *Our Common Future*, op. cit. See especially Chapter 3, 'The Role of the International Economy'.
17. See Lester R. Brown *et al*, *State of the World, 1988* (Washington, DC: Worldwatch Institute, 1988), p. 183. Also MacNeill, 'Strategies for Sustainable Economic Development', op. cit.
18. *Our Common Future*, op. cit. See especially Chapters 11 and 12.
19. Declaration of The Hague, The Hague, 11 March 1989, deposited with the Government of the Kingdom of the Netherlands.
20. Economic Declaration, Summit of the Arch, Paris, 16 July 1989.
21. Ibid.
22. Commonwealth Heads of Government Meeting, Kuala Lumpur, 'The Langkawi Declaration on Environment'. Commonwealth Secretariat (London, 1989).

A Note to Researchers

The International Development Research Centre (IDRC) has been given the original documents of the World Commission on Environment and Development. The Brundtland Collection consists of forty-three volumes of background papers and studies presented to the World Commission and the submissions and transcripts of the verbal testimonies given at the Commission's eight public hearings.

The materials in the collection can be accessed by computer through IDRC's Development Data Bases Service. Online searches of the Library's data base, BIBLIOL, allow easy access by author, title and subject. This service is free but users must cover their own telecommunications costs. Copies of individual papers on microfiche may be obtained free of charge. Microfiche sets of the complete collection are free to environmental institutions in the Third World. Canadian institutions can purchase sets directly from IDRC. Information about this collection and access to it is available from: IDRC Library, P.O. Box 8500, Ottawa K1G 3H9, Canada (tel (613) 598–0578: fax (613) 238–7230).

Institutions in other countries are able to purchase sets from the Centre for Our Common Future, which also publishes a *Brundtland Bulletin* which includes information on follow-up activities around the world. For information contact: Centre for Our Common Future, Palais Wilson, 52, rue des Paquis, CH-1201 Geneva, Switzerland (tel (022) 732 71 17: fax (022) 738 50 46).

Index